Energy Harvesting Systems

Tom J. Kaźmierski · Steve Beeby
Editors

Energy Harvesting Systems

Principles, Modeling and Applications

Springer

Editors
Tom J. Kaźmierski
School of Electronics and Computer Science
University of Southampton
Southampton, SO17 1BJ, UK
tjk@ecs.soton.ac.uk

Steve Beeby
School of Electronics and Computer Science
University of Southampton
Southampton, SO17 1BJ, UK
spb@ecs.soton.ac.uk

ISBN 978-1-4899-9878-1 ISBN 978-1-4419-7566-9 (eBook)
DOI 10.1007/978-1-4419-7566-9
Springer New York Dordrecht Heidelberg London

Printed on acid-free paper

Springer is part of Springer Science+Business Media (www.springer.com)

Preface

Energy harvesting is the conversion of ambient energy present in the environment into electrical energy. It is identical in principle to large-scale renewable energy generation, for example, solar or wind power, but very different in scale. While large-scale power generation is concerned with megawatts of power, energy harvesting typically refers to micro- to milli-watts, i.e. much smaller power generation systems. The development of energy harvesting has been driven by the proliferation of autonomous wireless electronic systems. A classic example of such systems are wireless sensor nodes which combine together to form wireless sensor networks. Each sensor node typically comprises a sensor, processing electronics, wireless communications, and power supply. Since the system is by definition wireless and cannot be plugged into a mains supply, power has to be provided locally. Typically such a local power supply is provided a battery which on the face of it is convenient and low cost. However, batteries contain a finite supply of energy and require periodic replacement or recharging. This may be fine in individual deployments but across a wireless network containing a multitude of nodes batteries are clearly not attractive. Furthermore, the need to replace batteries means the wireless system has to be accessible which may not be possible or may compromise performance. Finally, there are environmental concerns about disposing of batteries. Energy harvesting was developed, therefore, as a method for replacing or augmenting batteries. By converting ambient energy in the environment, the energy harvester can provide the required electrical power for the lifetime of the wireless system which is also free to be embedded or placed wherever it is best suited to perform its function. Energy harvesting typically exploit kinetic, thermal, solar sources, or electromagnetic radiation sources. Kinetic energy harvesting converts movement, often in the form of vibrations, into electrical energy. Thermal gradients can be exploited by using thermoelectric generators while solar energy is harvested using photovoltaics. Electromagnetic radiation can capture the energy from radio waves but unless this energy is specifically broadcast, power levels are typically very low. The challenges for energy harvesting are to maximise the available electrical power from the ambient energy found in the application environment. Vibration energy harvesters, for example, need to be tuned to match characteristic frequencies found in the environment which often means bespoke generator designs are required for different applications. It would be much better if such generators were adaptable and able to cope with a range of frequencies. The energy conversion process does not stop with the generator; typically power conditioning electronics are also required to provide the electrical power in form acceptable to the system electronics. The design of the conditioning electronics often has an impact on the performance of the generator and therefore energy harvesting systems should be designed in a holistic manner considering all the essential blocks as a whole. This book addresses these challenges and describes the approaches that can be taken to overcome them. The first chapter provides a comprehensive introduction to operating principles of kinetic micro-generators and associated electronics with emphasis on adaptive kinetic energy harvesting. Kinetic energy harvesters, also known as vibration power generators, are typically, although not exclusively, inertial spring-mass systems where electrical power is extracted by employing one or a combination of different transduction mechanisms. As most vibration power generators are resonant systems, they generate maximum power when the resonant frequency of the generator matches the ambient vibration frequency. Adaptive generators try to minimise the difference between these two frequencies in order to maximise the amount of generated power. The chapter outlines extensively recent developments in adaptive kinetic energy harvesting and presents achievable improvements in the operating frequency range of such generators. The second chapter is devoted to design automation aspects of energy harvester systems. It presents an automated energy harvester design flow which is based on a single HDL software platform that can be used to model, simulate, configure and optimise a complete mixed physical-domain energy harvester system which includes the micro-generator, voltage booster, storage element and load. State-of-the art accurate hardware description language (HDL) modelling techniques for kinetic energy harvesters and their experimental validation are presented and discussed. Measurements have validated both the accuracy of HDL-based modelling and the efficiency of the automated design flow which can improve the amount of harvested energy in a typical system by 75%. The third chapter focuses on the power analysis and power harvesting in wireless sensor networks. Specifically, it gives an overview of power analysis simulation techniques and presents related tools and methodologies. The chapter also describes extension libraries for power analysis and models of energy harvester-based wireless sensors in SystemC and SystemC AMS complete with examples and simulation results. The final chapter presents a major industrial application of energy harvesting: the tyre pressure monitoring system (TPMS) recently developed at Infineon Technologies. Existing TPMS designs, which are based on batteries, suffer from difficulties in the energy budget management. The main challenge is to ensure a reliable RF data link from the sensor in the tyre to the receiver in the car due to the low energy available. Problems are aggravated when the car is used in extreme weather conditions, especially in winter, as batteries frequently fail in low temperatures. The chapter presents the advantages of Infineons revolutionary design which is based on an electrostatic vibration harvester. The design is highly miniaturised, with a volume less than 1 cm^3 including the power supply and is embedded in the tyre. Results obtained from this case study provide a significant step towards intelligent tyres, which would be able to measure and report additional technical parameters for further enhancement of road safety. We hope that the reader of this book will gain a valuable insight into the state-of-the-art design techniques for autonomous wireless sensors powered by kinetic energy harvesters. The potential for electronic systems using various forms of "free energy", such as kinetic, thermal, solar, RF and others, will continue to inspire researchers and engineers. In near future we will no doubt see new energy harvester designs, use of new materials, as well as innovative power management circuits and new solutions to energy storage. Energy harvester applications will benefit from further evolution in decreasing energy consumption due to novel circuit designs and the scaling down of nano-devices. New wireless communication techniques will also contribute to the reduction of energy consumption.

Southampton, UK

Tom J. Kaźmierski
Steve Beeby

Contents

Contributors

Steve Beeby School of Electronics and Computer Science, University of Southampton, Southampton UK, spb@ecs.soton.ac.uk

Christoph Grimm Institute of Computer Technology, 1040 Vienna, Austria, grimm@ict.tuwien.ac.at

Jan Haase Vienna University of Technology, Institute of Computer Technology, Gusshausstrasse 27-29, 1040 Vienna, Austria, haase@ict.tuwien.ac.at

Thomas Herndl Infineon Technologies Austria AG, Vienna, Austria, thomas.herndl@infineon.com

Tom J. Kaźmierski School of Electronics and Computer Science, University of Southampton, Southampton, UK, tjk@ecs.soton.ac.uk

Jiong Ou Institute of Computer Technology, 1040 Vienna, Austria, ou@ict.tuwien.ac.at

Leran Wang School of Electronics and Computer Science, University of Southampton, Southampton, UK, lw04r@ecs.soton.ac.uk

Joseph Wenninger Institute of Computer Technology, 1040 Vienna, Austria wenninger@ict.tuwien.ac.at

Dibin Zhu School of Electronics and Computer Science, University of Southampton, Southampton, UK, dz@ecs.soton.ac.uk

Chapter 1
Kinetic Energy Harvesting

Dibin Zhu and Steve Beeby

Abstract This chapter introduces principles of normal kinetic energy harvesting and adaptive kinetic energy harvesting. Kinetic energy harvesters, also known as vibration power generators, are typically, although not exclusively, inertial spring-mass systems. Electrical power is extracted by employing one or a combination of different transduction mechanisms. Main transduction mechanisms are piezoelectric, electromagnetic and electrostatic. As most vibration power generators are resonant systems, they generate maximum power when the resonant frequency of the generator matches ambient vibration frequency. Any difference between these two frequencies can result in a significant decrease in generated power. Recent development in adaptive kinetic energy harvesting increases the operating frequency range of such generators. Possible solutions include tuning resonant frequency of the generator and widening the bandwidth of the generator. In this chapter, principles and operating strategies for adaptive kinetic energy harvesters will be presented and compared.

Keywords Adaptive energy harvesting · Frequency tuning · Wider frequency range · Vibration energy harvesting

1.1 Introduction

Mechanical energy can be found almost anywhere that wireless sensor networks (WSN) may potentially be deployed, which makes converting mechanical energy from ambient vibration into electrical energy an attractive approach for powering wireless sensors. The source of mechanical energy can be a moving human body or a vibrating structure. The frequency of the mechanical excitation depends on the source: less than 10 Hz for human movements and over 30 Hz for machinery

D. Zhu (✉)
School of Electronics and Computer Science, University of Southampton, Southampton, UK
e-mail: dz@ecs.soton.ac.uk

T.J. Kaźmierski, S. Beeby (eds.), *Energy Harvesting Systems*,
DOI 10.1007/978-1-4419-7566-9_1, © Springer Science+Business Media, LLC 2011

vibrations [1]. Such devices are known as kinetic energy harvesters or vibration power generators [2].

In practical machine-based applications, vibration levels can be very low (<1 m s^{-2}) at frequencies that often correspond to the frequency of the mains electricity powering the plant (e.g. 50 or 60 Hz or harmonics). Such low levels of vibration equate to amplitudes of vibration that are in the order of a few microns and the only way to extract mechanical energy in this case is to use an inertial generator that resonates at a characteristic frequency. The limitation to this approach is that the generator is, by definition, designed to work at a single frequency. A high Q-resonance means very limited practical bandwidths over which energy can be harvested. If the resonant frequency does not match the ambient vibration frequency, output power of the generator drops significantly.

Adaptive kinetic energy harvesters [3] are developed to increase the operational frequency range of vibration energy harvesters thus addressing the bandwidth limitation. Adaptive kinetic energy employs certain mechanisms that can either adjust, or tune, the resonant frequency of a single generator so that it matches the frequency of the ambient vibration at all times or widen the bandwidth of the generator. Resonant frequency tuning can be achieved by changing the mechanical characteristics of the structure or electrical load on the generator. In addition, widening the bandwidth of the generator can be achieved by, for example, employing an array of structures each with a different resonant frequency, an amplitude limiter, coupled oscillators, non-linear (e.g. magnetic) springs, bi-stable structures or a large inertial mass (large device size) with a high degree of damping.

In Section 1.2, principles of kinetic energy harvesting is introduced. In Section 1.3, classification of transduction mechanisms and principle of each transducer are described. A wide range of reported kinetic energy harvesters are summarized according to their transducers. Advantages and disadvantages of each transducer have been listed and compared. In Section 1.4, a brief introduction of adaptive kinetic energy harvesting is given. Section 1.5 describes the theory behind resonant frequency tuning strategies and suggests criteria for evaluating tuning mechanisms. Principles of the two tuning methods, i.e. mechanical and electrical tuning, are introduced and examples of these methods are studied. Section 1.6 presents principles of strategies to widen bandwidth of the kinetic energy harvesters and contains examples of all strategies. Section 1.7 compares different strategies for adaptive kinetic energy harvesting and Section 1.8 summarizes the chapter.

1.2 Principles of Kinetic Energy Harvesting

Inertial-based kinetic energy harvesters are modelled as second-order, spring-mass systems. The generic model of kinetic energy harvesters was first developed by Williams and Yates [4]. Figure 1.1 shows a generic model of such a generator, which consists of a seismic mass, m, and a spring with the spring constant of k. When the generator vibrates, the mass moves out of phase with the generator housing. There

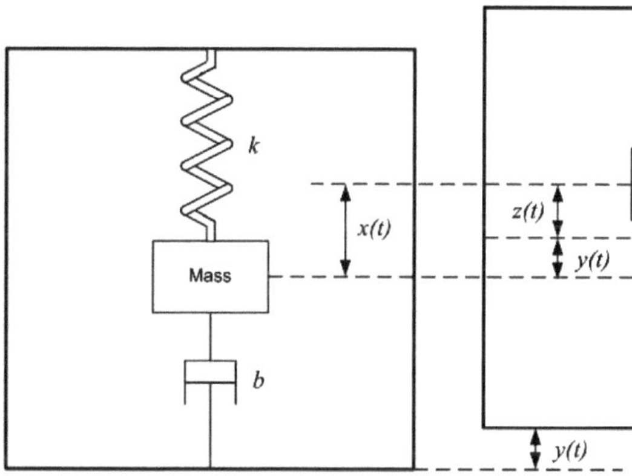

Fig. 1.1 Generic model of kinetic energy harvesters

is a relative movement between the mass and the housing. This displacement is sinusoidal in amplitude and can drive a suitable transducer to generate electrical energy. b is the damping coefficient that consists of mechanically induced damping (parasitic damping) coefficient b_m and electrically induced damping coefficient b_e, i.e. $b = b_m + b_e$. $y(t)$ is the displacement of the generator housing and $z(t)$ is the relative motion of the mass with respect to the housing. For a sinusoidal excitation, $y(t)$ can be written as $y(t) = Y \sin \omega t$, where Y is the amplitude of vibration and ω is the angular frequency of vibration.

The transduction mechanism itself can generate electricity by exploiting the mechanical strain or relative displacement occurring within the system. The strain effect utilizes the deformation within the mechanical system and typically employs active materials (e.g. piezoelectric). In the case of relative displacement, either the velocity or position can be coupled to a transduction mechanism. Velocity is typically associated with electromagnetic transduction while relative position is associated with electrostatic transduction. Each transduction mechanism exhibits different damping characteristics and this should be taken into consideration while modelling the generators. Thermomechanical system can be increased in complexity, for example, by including a hydraulic system to magnify amplitudes or forces, or couple linear displacements into rotary generators. Details of these transducers will be given in Section 1.3.

1.2.1 Transfer Function

For the analysis, it is assumed that the mass of the vibration source is much greater than the mass of seismic mass in the generator and the vibration source is unaffected by the movement of the generator. Then the differential equation of the movement of the mass with respect to the generator housing from the dynamic forces on the mass can be derived as follows:

$$m \cdot \frac{d^2 z(t)}{dt^2} + b \cdot \frac{dz(t)}{dt} + k \cdot z(t) = -m \cdot \frac{d^2 y(t)}{dt^2} \tag{1.1}$$

which can be written in the form after the Laplace transform as

$$m \cdot s^2 \cdot z(s) + b \cdot s \cdot z(s) + k \cdot s \cdot z(s) = -m \cdot a(s) \tag{1.2}$$

where $a(s)$ is the Laplace expression of the acceleration of the vibration, $a(t)$, which is given by

$$a(t) = \frac{d^2 y(t)}{dt^2} \tag{1.3}$$

Thus, the transfer function of a vibration-based micro-generator is

$$\frac{z(s)}{a(s)} = \frac{1}{s^2 + \frac{b}{m}s + \frac{k}{m}} = \frac{1}{s^2 + \frac{\omega_r}{Q}s + \omega_r^2} \tag{1.4}$$

where $Q = \frac{\sqrt{km}}{b}$ is the quality factor and $\omega_r = \sqrt{k/m}$ is the resonant frequency.

1.2.2 Equivalent Circuit

An equivalent electrical circuit for a kinetic energy harvester can be found from Eq. (1.4), which, when rearranged, gives

$$-m \cdot a(s) = s \cdot Z(s) \left(ms + b + \frac{k}{s} \right) \tag{1.5}$$

Equation (1.5) can be rewritten as

$$-I(s) = E(s) \left(sC + \frac{1}{R} + \frac{1}{sL} \right) \tag{1.6}$$

where $I(s) = m \cdot a(s)$, $E(s) = s \cdot Z(s)$, $C = m$, $R = \frac{1}{b}$, $L = \frac{1}{k}$. Based on Eq. (1.6), an equivalent electrical circuit can be built as shown in Fig. 1.2.

1.2.3 Damping in Kinetic Energy Harvesters

As mentioned above, damping in kinetic energy harvesters consists of mechanically induced damping (parasitic damping) and electrically induced damping. The overall damping factor of the system, ζ_T, is given by

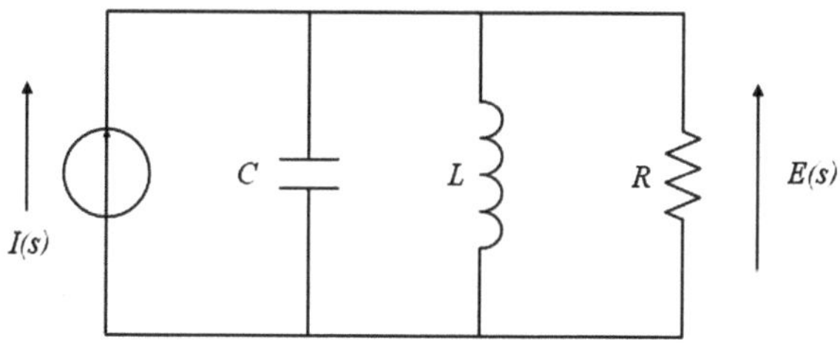

Fig. 1.2 Equivalent circuit of a kinetic energy harvester

$$\zeta_T = \frac{b}{2m\omega_r} = \frac{b_m + b_e}{2m\omega_r} = \zeta_m + \zeta_e \tag{1.7}$$

where $\zeta_m = \frac{b_m}{2m\omega_r}$ is the mechanically induced damping factor and $\zeta_e = \frac{b_e}{2m\omega_r}$ is the electrically induced damping factor.

Total quality factor (Q-factor) is a function of damping factor. The total Q-factor is given by

$$Q_T = \frac{1}{2\zeta_T} \tag{1.8}$$

This is the Q-factor when the generator is connected to the optimum load. The relation between total quality factor and the electrical and mechanical damping is given by

$$\frac{1}{Q_T} = \frac{1}{Q_{OC}} + \frac{1}{Q_e} \tag{1.9}$$

where $Q_{OC} = \frac{1}{2\zeta_m}$ is the open circuit Q-factor which reflects the mechanical damping. Q_e, which equals $\frac{1}{2\zeta_e}$, reflects performance of the transduction mechanism. It cannot be measured directly, but can be calculated using Eq. (1.9) once Q_T and Q_{OC} are measured.

1.2.4 Output Power of Kinetic Energy Harvesters

Assume that the input is a sinusoid excitation, i.e. $y(t) = \sin \omega t$. The solution to Eq. (1.1) is given by

$$z(t) = \frac{m\omega^2 Y}{k - m\omega^2 + j\omega b} \cdot \sin \omega t \tag{1.10}$$

or

$$z(t) = \frac{\omega^2}{\sqrt{(\omega_r^2 - \omega^2)^2 + (\frac{b\omega}{m})^2}} \cdot Y \sin(\omega t + \varphi) \tag{1.11}$$

where φ is the phase angle given by

$$\varphi = \tan^{-1}\left(\frac{b\omega}{k - \omega^2 m}\right) \tag{1.12}$$

The average power dissipated within the damper, i.e. the sum of the power extracted by the transduction mechanism and the power lost in mechanical damping is given by

$$P = b\left(\frac{dz(t)}{dt}\right)^2 \tag{1.13}$$

Equations (1.11) and (1.13) give the average power dissipated within the damper as follows:

$$P(\omega) = \frac{m\zeta_T Y^2 (\frac{\omega}{\omega_r})^3 \omega^3}{[1 - (\frac{\omega}{\omega_r})^2]^2 + (2\zeta_T \frac{\omega}{\omega_r})^2} \tag{1.14}$$

When the generator is at resonance, i.e. $\omega = \omega_r$, the power dissipation reaches maximum. The maximum dissipated power is

$$P = \frac{mY^2\omega_r^3}{4\zeta_T} \tag{1.15}$$

or

$$P = \frac{mY^2\omega_r^3}{4(\zeta_m + \zeta_e)} \tag{1.16}$$

The power dissipation is the sum of maximum electrical energy extracted by the transduction mechanism, P_e, and mechanical loss, P_m. P_e and P_m are as follows:

$$P_e = \frac{\zeta_e m Y^2 \omega_r^3}{4(\zeta_m + \zeta_e)} \tag{1.17}$$

$$P_m = \frac{\zeta_m m Y^2 \omega_r^3}{4(\zeta_m + \zeta_e)} \tag{1.18}$$

Maximum power conversion from mechanical domain to electrical domain occurs when $\zeta_e = \zeta_m$, i.e. damping arising from the electrical domain equals mechanical losses. Therefore, the maximum electrical power that can be extracted by the kinetic energy harvester, P_e, is given by

$$P_e = \frac{P}{2} = \frac{mY^2\omega_r^3}{16\zeta_m} \tag{1.19}$$

Since the peak acceleration of the base, a, is given by $a = Y\omega^2$, Eq. (1.19) can be rewritten as

$$P_e = \frac{ma^2}{16\omega_r \cdot \zeta_m} \tag{1.20}$$

As the open circuit Q-factor, $Q_{OC} = \frac{1}{2\zeta_m}$, Eq. (1.20) can be written as

$$P_e = \frac{ma^2}{8\omega_r} \cdot Q_{OC} \tag{1.21}$$

It is found via Eq. (1.21) that the maximum power delivered to the electrical domain is inversely proportional to the damping factor, i.e. proportional to the Q-factor. Hence, when designing a vibration-based micro-generator to achieve maximum power output, it is important to design the generator with a high Q-factor (i.e. low damping factor) and make the generator work at its resonant frequency. Figure 1.3 shows an example of the power spectrum of a vibration-based micro-generator of resonant frequency 50 Hz with various Q-factors and damping factors. It can be seen that, for generators with a high Q-factor (i.e. low damping factor), the output power drops significantly if the frequency of operation is away from the generators resonance. When the Q-factor is lower (i.e. damping factor is higher), the peak output power decreases while the bandwidth of the generator increases and the device becomes less sensitive to frequency shifts at the expense of lower maximum generated power. In addition, since the output power is inversely proportional to the resonant frequency of the generator for a given acceleration, it is generally preferable to operate at the lowest available fundamental frequency. This is compounded by practical observations that acceleration levels associated with environmental vibrations tend to reduce with increasing frequency. Application vibration spectra should be carefully studied before designing the generator in order to correctly identify the frequency of operation given the design constraints on generator size and maximum permissible $z(t)$. Furthermore, the mass of the mechanical structure should be maximized within the given size constraints in order to maximize the electrical power output. It should also be noted that the energy delivered to the electrical domain will not necessarily all be usefully harvested (e.g. coil losses).

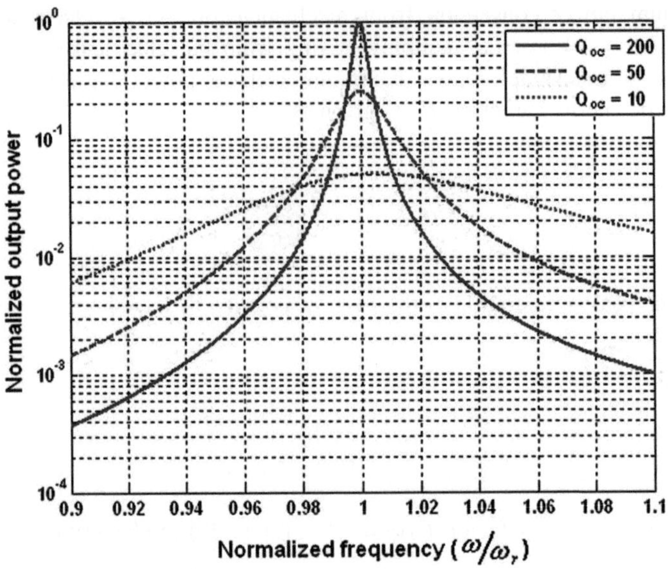

Fig. 1.3 Power spectrum of a kinetic energy harvester with various Q-factors

1.3 Transduction Mechanisms

In kinetic energy harvesting, a particular transduction mechanism such as electromagnetic [5], electrostatic [6] and piezoelectric [7] is used to extract electrical energy from motion. The generator also requires a mechanical system to couple environmental displacements to the transduction mechanism. This mechanical system has to be designed to maximize the coupling between the mechanical energy source and the transduction mechanism. Most vibration-based micro-generators are single degree of freedom second-order spring-mass system consisting of an inertial frame that transmits the vibration to a suspended inertial mass to produce a relative displacement or cause mechanical strain. The transduction mechanism can then generate electrical energy by exploiting the relative displacement or strain.

1.3.1 Electromagnetic (EM) Generators

Electromagnetic induction was discovered by Michael Faraday in 1831. Faraday's law of electromagnetic induction states that an electrical current will be induced in any closed circuit when the magnetic flux through a surface bounded by the conductor changes. This applies whether the field itself changes in strength or the conductor is moved through it. In an electromagnetic generator, permanent magnets are used to produce strong magnetic field and a coil is used as the conductor. Either the permanent magnet or the coil is fixed to the frame while the other is attached to the inertial mass. The relative displacement caused by the vibration makes the transduction mechanism work and generate electrical energy. The induced voltage, also known as electromotive force (e.m.f), across the coil is proportional to the strength of the magnetic field, the velocity of the relative motion and the number of turns of the coil. An electromagnetic generator is characterized by high output current level at the expense of low voltages. Figure 1.4 shows two commonly seen examples of electromagnetic generators.

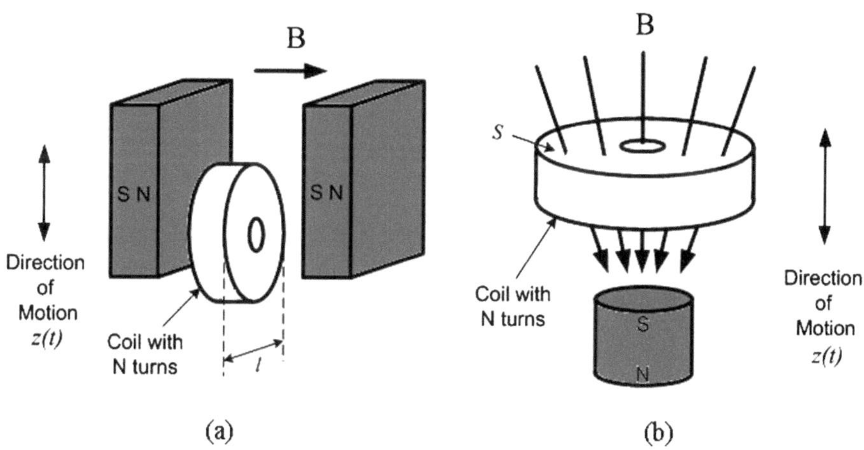

Fig. 1.4 Electromagnetic generators

For the case in Fig. 1.4a, the magnetic field is uniform. The magnetic field cut by the coil varies with the relative displacement between magnets and the coil. In this case, the induced electromotive force is given by

$$\text{e.m.f.} = -N \cdot l \cdot B \cdot \frac{dz}{dt} \tag{1.22}$$

where N is the number of turns of the coil, l is the effective length of the coil, B is the flux density going through the coil and dz/dt is the relative velocity between the magnets and the coil.

For the case in Fig. 1.4b, the magnetic field varies with the distance apart from the magnet. The induced electromotive force is given by

$$\text{e.m.f.} = -N \cdot S \cdot \frac{dB}{dz} \cdot \frac{dz}{dt} \tag{1.23}$$

where S is the effective area of the coil and dB/dz is the gradient of the magnetic flux density along the direction of relative motion between magnets and the coil.

In both cases, the induced e.m.f. is a function of velocity of relative movement $z(t)$. Therefore, both expressions can be expressed by

$$\text{e.m.f.} = \kappa \cdot \frac{dz}{dt} \tag{1.24}$$

where κ is the electromagnetic coupling factor which equals $-N \cdot l \cdot B$ and $-N \cdot S \cdot \frac{dB}{dz}$ in both cases, respectively. It represents the change in coupled flux per unit displacement.

Figure 1.5 shows a circuit representation of an electromagnetic generator with a resistive load, R_L. The relation between the current through the load and the induced e.m.f. is given by

$$\text{e.m.f.} + i \cdot (R_L + R_c) + L_c \frac{di}{dt} = 0 \tag{1.25}$$

Electromagnetic generators perform better in macro-scale than in micro-scale [8]. Particularly, generators integrated with MEMS with electroplated coils and magnets may not be able to produce useful power levels due to poor electromagnetic coupling.

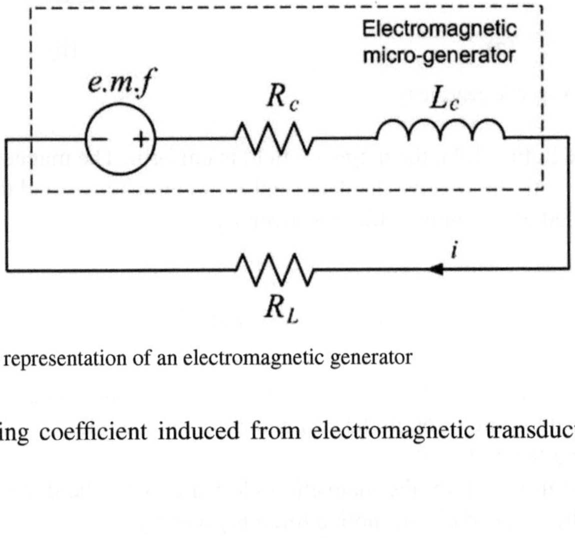

Fig. 1.5 Circuit representation of an electromagnetic generator

The damping coefficient induced from electromagnetic transduction, b_e, is as follows:

$$b_e = \frac{\kappa^2}{R_L + R_c + j\omega L_c} \tag{1.26}$$

where R_L and R_c are resistances of the load and coil, respectively. L_c is the inductance of the coil.

For a micro-generator that works at low resonant frequencies, the inductive impedance of the coil is much lower than its resistive impedance. Hence, the inductive impedance can be ignored in this case. Thus, b_e, can be simplified to

$$b_e = \frac{\kappa^2}{R_L + R_c} \tag{1.27}$$

The electrically induced damping factor, ζ_e, is

$$\zeta_e = \frac{\kappa^2}{2m\omega(R_L + R_c)} \tag{1.28}$$

Equation (1.28) shows that R_L can be used to adjust b_e to match b_m and therefore maximize output power, although this must be done with the coil parameters in mind. It can be shown that the optimum load resistance can be found from Eq. (1.29) and maximum average power delivered to the load can be found from Eq. (1.30) [9]:

$$R_L = R_c + \frac{\kappa^2}{b_m} \tag{1.29}$$

$$P_e = \frac{ma^2}{16\zeta_m\omega_r(1 - \frac{R_c}{R_L})} \tag{1.30}$$

Table 1.1 lists some reported electromagnetic generators with their main characteristics.

Table 1.1 Summary of electromagnetic kinetic energy harvesters

Reference	f (Hz)	Excitation level (m s^2)	Mass (g)	Volume (mm^3)	P (μW)	Power density (μW mm^3)	Structure material
Williams et al. [10]	4400	382	0.0023	5.4	0.3	0.0556	GaAs Polyimide[b]
Ching et al. [11]	110	95.5	N/A	1000	830	0.83	Copper[c]
Glynne-Jones et al. [12]	322	2.7	N/A	840	180	0.214	Steel[c]
Koukharenko et al. [13]	1615	3.92	N/A	100	0.104	0.00104	Silicon[b]
Saha et al. [14]	84	7.8	25	800[a]	3500	4.375	Copper[c]
Beeby et al. [15]	52	0.589	0.66	150	46	0.307	BeCu[c]
Klahand et al. [16]	25	N/A	15.6[a]	2000[a]	3.97	0.00199	Styrene[b]
Torah et al. [17]	50	0.589	N/A	570	58	0.102	BeCu[c]
Wang et al. [18]	280	10	N/A	315	17.2	0.055	Nickel[b]

[a] Estimated or extrapolated from data in reference
[b] Micro-scale
[c] Macro-scale

1.3.2 Piezoelectric (PZ) Generators

The piezoelectric effect was discovered by Pierre and Jacques Curie in 1880. It is the ability of some materials (notably crystals and certain ceramics) to generate an electric potential in response to applied mechanical stress. The electrical polarization is proportional to the applied strain. This is the piezoelectric effect used for mechanical to electrical energy conversion. Commonly used materials for piezoelectric power generation are PZT, PVDF [19] and macro-fibre composite (MFC) [20].

Piezoelectric generators typically work in either 33 mode (Fig. 1.6a) or 31 mode (Fig. 1.6b). In the 33 mode, a force is applied in the same direction as the poling direction, such as the compression of a piezoelectric block that has electrodes on its top and bottom surfaces. In the 31 mode, a lateral force is applied in the direction perpendicular to the poling direction, an example of which is a bending beam that has electrodes on its top and bottom surfaces. Generally, the 31 mode has been the most commonly used coupling mode although the 31 mode has a lower coupling coefficient than the 33 mode [19]. Common energy harvesting structures such as cantilevers or double-clamped beam typically work in the 31 mode because the lateral stress on the beam surface is easily coupled to piezoelectric materials deposited onto the beam.

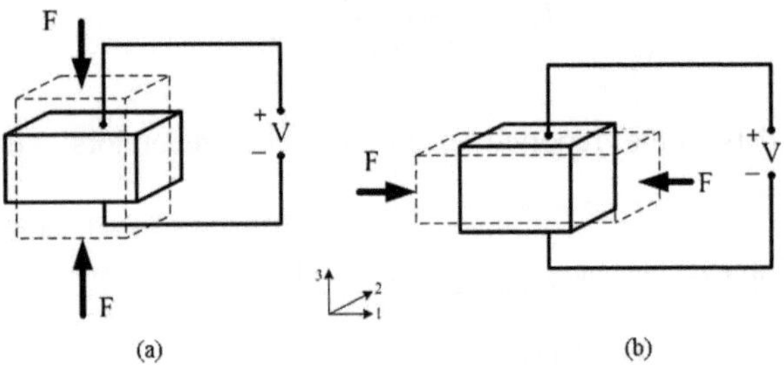

Fig. 1.6 Piezoelectric generators: (**a**) 33 mode and (**b**) 31 mode

The constitutive equations for a piezoelectric material are given by

$$\delta = \frac{\sigma}{Y} + d \cdot E \tag{1.31}$$

$$D = \varepsilon \cdot E + d \cdot \sigma \tag{1.32}$$

where δ is mechanical strain, σ is mechanical stress, Y is Young's modulus of the material, d is the piezoelectric strain coefficient, E is the electric field, D is the electrical displacement (charge density) and ε is the dielectric constant of the piezoelectric material.

Figure 1.7 shows a circuit representation of a piezoelectric generator with a resistive load, R_L. C is the capacitance between two electrodes and R_s is the resistance

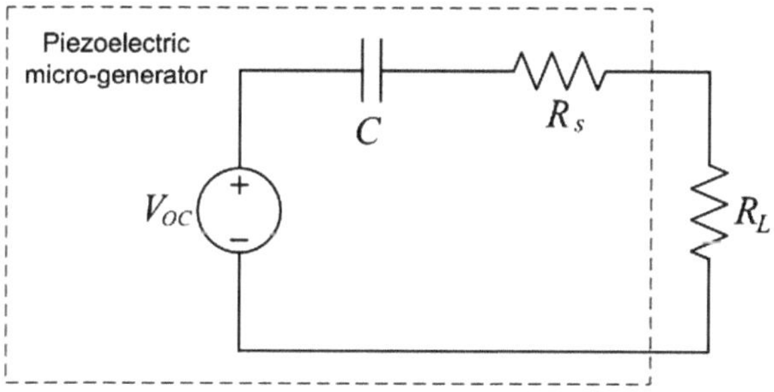

Fig. 1.7 Circuit representation of a piezoelectric generator

of the piezoelectric material. The voltage source, V_{OC}, is the open circuit voltage resulting from Eq. (1.32) when the electrical displacement is zero. It is given by

$$V_{OC} = -\frac{d \cdot t}{\varepsilon} \cdot \sigma \tag{1.33}$$

where t is the thickness of the piezoelectric material.

An expression for the piezoelectric damping coefficient is [21]

$$b_e = \frac{2m\omega_r^2\kappa^2}{2\sqrt{\omega_r^2 + \frac{1}{R_L C_L}}} \tag{1.34}$$

where κ is the piezoelectric material electromechanical coupling factor and C_L is the load capacitance. Again R_L can be used to optimize and the optimum value can be found from Eq. (1.35) and as stated previously, maximum power occurs when ζ_e equals ζ_m:

$$R_{opt} = \frac{1}{\omega_r C} \frac{2\zeta_m}{\sqrt{4\zeta_m^2 + \kappa^4}} \tag{1.35}$$

The maximum power is [21]

$$P_{max} = \frac{1}{\omega_r^2} \frac{R_L C^2 (\frac{2Ydtb^*}{\varepsilon})^2}{(4\zeta_m^2 + \kappa^4)(R_L C\omega_r)^2 + 4\zeta_m\kappa^2(R_L C\omega_r) + 2\zeta_m^2} a^2 \tag{1.36}$$

where b^* is a constant related to dimensions of the piezoelectric generator and a is the vibration acceleration.

Table 1.2 lists properties of some common piezoelectric materials. Output power of piezoelectric generators using different piezoelectric materials is compared in

Table 1.2 Coefficients of common piezoelectric materials [2, 22]

Material	PZT-5H	PZT-5A	$BaTiO_3$	PVDF
d_{31} ($\times 10^{-12} CN^{-1}$)	−274	−171	78	23
Young's modulus (GPa)	50	50	67	2
Relative permittivity ($\varepsilon/\varepsilon_0$)	3400	1700	1700	12

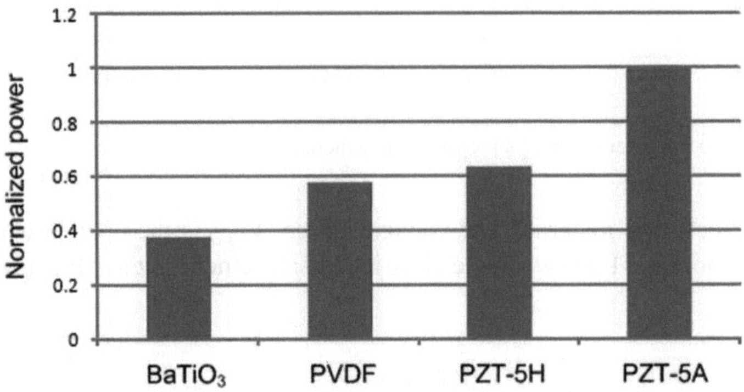

Fig. 1.8 Comparison of output power of piezoelectric generator using different piezoelectric materials

Fig. 1.8. These generators have the same dimensions. It is found that with the same dimensions, the generator using PZT-5A has the most amount of output power.

Piezoelectric generators have the simplest structure among the three transducers and they can produce appropriate voltages for electronic devices. However, the mechanical properties of the piezoelectric material may limit overall performance and lifespan of the generator. Although piezoelectric thin films can be integrated into a MEMS fabrication process, the piezoelectric coupling is greatly reduced. Therefore, the potential for integration with microelectronics is less than that for electrostatic micro-generators which will be presented in the next section.

Table 1.3 lists some reported piezoelectric generators with their main characteristics.

1.3.3 Electrostatic (ES) Generators

The basis of electrostatic generator is the variable capacitor. The variable capacitance structure is driven by mechanical vibrations. The capacitance varies between maximum and minimum values. If the charge on the capacitor is constrained, charge will move from the capacitor to a storage device or to the load as the capacitance decreases. Thus, mechanical energy is converted to electrical energy. Electrostatic generators can be classified into three types, i.e. in-plane overlap (Fig. 1.9a) which varies the overlap area between electrode fingers, in-plane gap closing (Fig. 1.9b) which varies the gap between electrode fingers and out-of-plane gap closing (Fig. 1.9c) which varies the gap between two large electrode plates [6].

Table 1.3 Summary of piezoelectric kinetic energy harvesters

Reference	f (Hz)	Excitation level (m s^2)	Mass (g)	Volume (mm^3)	P (μW)	Power density (μW mm^3)	Structure	Piezoelectric material
White et al. [23]	80	2.3	0.8	125	2.1	0.0168	Unimorph	Screen-printed PZT
Roundy et al. [1]	120	2.5	9.2	1000	375	0.375	Bimorph	PZT
Lu et al. [24]	7000	N/A	N/A	N/A	1600[a]	N/A	Bimorph	PZT-PIC255
Jeon et al. [25]	13.9	106	N/A	0.027[b]	1	37.04	Unimorph	PZT
Fang et al. [26]	608	9.8	0.0016[b]	0.6[b]	2.16	3.6	Unimorph	PZT
Marzencki et al. [27]	1500	3.92	0.0009[b]	5	0.03	0.006	Unimorph	AlN
Jeong et al. [28]	120	0.98	N/A	N/A	500	22	Bimorph	PMNZT
Kok et al. [29]	230	9.8	N/A	N/A	0.27	N/A	Free-standing unimorph	PZT
Shen et al. [30]	462.5	19.6	N/A	N/A	2.15	3.3	Unimorph	PZT
Zhu et al. [31]	67	4	2.8	987	240	0.24	Bimorph	Screen-printed PZT

[a] Simulation result

[b] Extrapolated from data in reference

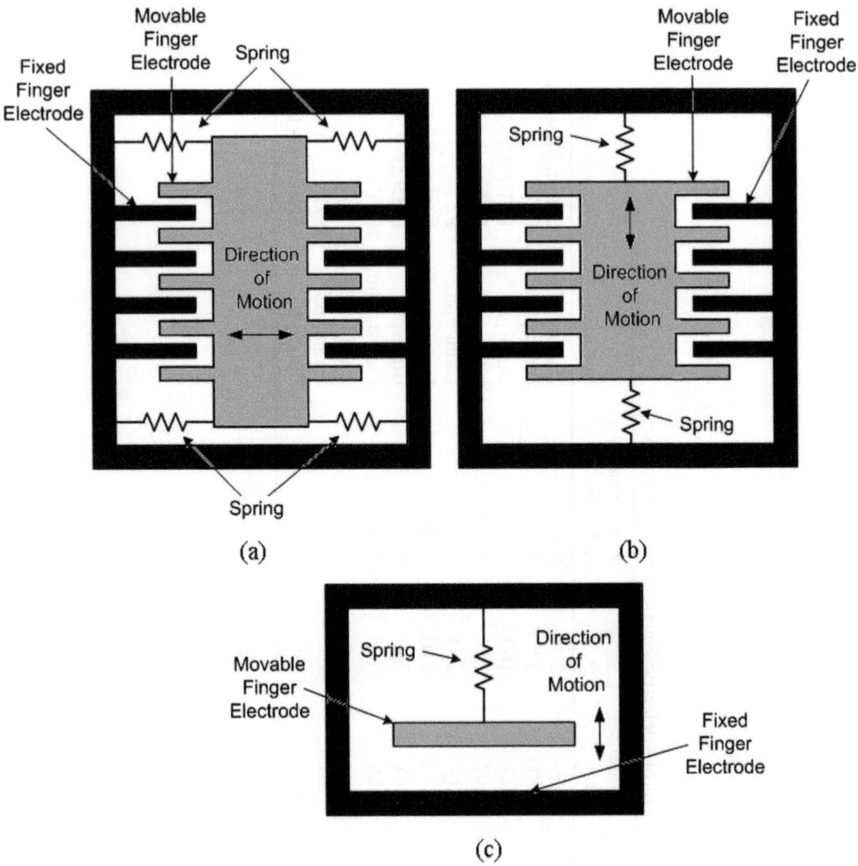

Fig. 1.9 Electrostatic generators: (**a**) in-plane overlap; (**b**) in-plane gap closing; and (**c**) out-of-plane gap closing

These three types can be operated either in charge-constrained or voltage-constrained cycles. Generally, generators working in voltage-constrained cycles provide more energy than generators in charge-constrained cycles. However, by incorporating a capacitor in parallel with the energy harvesting capacitor, the energy from the charge-constrained system can approach that of the voltage-constrained system as the parallel capacitance approaches infinity. This parallel capacitor effectively constrains the voltage on the energy harvesting capacitor [32].

A simplified circuit for an electrostatic generator using charge-constrained conversion is shown in Fig. 1.10. V_{in} is a pre-charged reservoir, which could be a capacitor or a rechargeable battery. C_v is a variable capacitor, which is one of the three types mentioned above. C_{par} is the parasitic capacitance associated with the variable capacitor structure and any interconnections, which limits the maximum voltage. C_L is the storage capacitor or any kind of load.

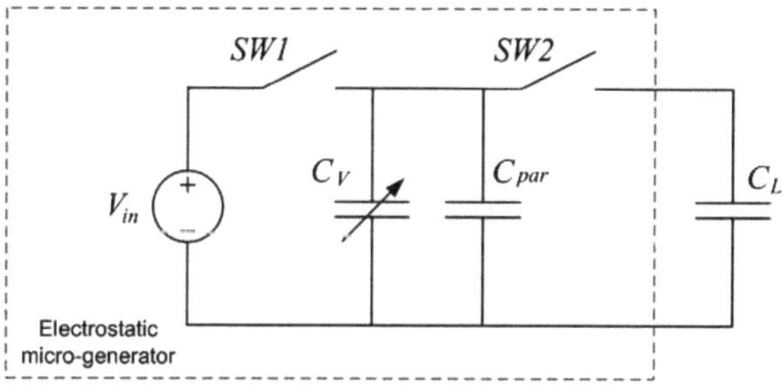

Fig. 1.10 Circuit representation for an electrostatic generator

The maximum voltage across the load is given by:

$$V_{\max} = \frac{C_{\max} + C_{\text{par}}}{C_{\min} + C_{\text{par}}} V_{\text{in}} \tag{1.37}$$

The energy dissipated within the damper, and therefore the power, is given by the force distance product shown in Eq. (1.37) [33]:

$$P = \frac{4YF\omega\omega_c^2}{2\pi}\sqrt{\frac{1}{1-\omega_c^2} - \left(\frac{F}{mY\omega^2\omega_c}U\right)^2} \tag{1.38}$$

where F is the damping force and Y is the displacement of the frame, $\omega_c = \omega/\omega_r$ and $U = \frac{\sin(\pi/\omega_c)}{1+\cos\pi/\omega_c}$.

The optimum damping force is given by

$$F_{\text{opt}} = \frac{Y\omega^2 m}{\sqrt{2}} \frac{\omega_c}{|(1-\omega_c^2)U|} \tag{1.39}$$

An electrostatic generator can be easily realized in MEMS version. Since the fabrication process of electrostatic generators is similar to that of VLSI, electrostatic generators can be assembled with VLSI without difficulties. Unfortunately, electrostatic generators require an initial polarizing voltage or charge. The output impedance of the devices is often very high, which makes them less suitable as a power supply. However, they can be used to charge a battery, in which case, electrostatic generators can use electrets to provide the initial charge.

Table 1.4 lists some reported electrostatic generators with their main characteristics.

Table 1.4 Summary of electrostatic kinetic energy harvesters

Reference	f (Hz)	Excitation level (m s^2)	Mass (g)	Volume (mm^3)	P (μW)	Power density (μW mm^3)	Type
Meninger et al. [34]	2520	N/A	N/A	75	8	0.11	IPO
Tashiro et al. [35]	6	1	780	N/A	36	N/A	OP
Mitcheson et al. [36]	30	50	0.1	750	3.7	0.0049	N/A
Arakawa et al. [37]	10	3.9	N/A	800	6	0.0075	IPO
Despesse et al. [38]	50	8.8	104	1800	1052	0.584	IPGC
Kuehne et al. [39]	1000	1.96	N/A	N/A	4.28	0.079	IPO
Yen et al. [40]	1560	82.32	N/A	N/A	1.8	N/A	OP
Sterken et al. [41]	500	9.8	N/A	N/A	5	N/A	OP
Lo and Tai [42]	50	576.6	54	50,000	17.98	0.00036	OP
Hoffmann et al. [43]	1300–1500	127.4	642e-6	N/A	3.5	N/A	IPO
Naruse et al. [44]	2	3.92	N/A	N/A	40	N/A	IPGC

IPO in-plane overlap, *IPGC* in-plane gap closing, *OP* out-of-plane

1.3.4 Other Transduction Mechanisms

Magnetostrictive materials are also used to extract electrical energy from ambient vibration. These materials deform when placed in a magnetic field and it can induce changes in magnetic field when it is strained. Magnetostrictive materials are generally used in piezoelectric-magnetostrictive composites. Such composites were originally used in magnetic field sensors and have recently been adopted in energy harvesting.

Huang et al. [45] reported two energy harvesting devices based on a Terfenol-D/PZT/Terfenol-D composite. Their device produced 1.2 mW of power when excited at 5 m s^{-2} at 30 Hz. Recently, Wang and Yuan [46] reported a new vibration energy harvester based on magnetostrictive material, Metglas 2605SC, with electromagnetic pickup. Experimentally, the maximum output power and power density on the load resistor can reach 200 μW and 900 μW cm^{-3}, respectively, at a low frequency of 58 Hz. For a working prototype under a vibration with resonance frequency of 1.1 kHz and peak acceleration of 8.06 m s^{-2}, the average power and power density during charging the ultracapacitor can achieve 576 μW and 606 μW cm^{-3}, respectively. Dai et al. [47] reported an energy harvester that converts ambient mechanical vibration into electrical energy employing the Terfenol-D/PZT/Terfenol-D laminate magnetoelectric (ME) transducer. The harvester uses four magnets arranged on the free end of a cantilever beam. The magnets produce a concentrated flux gradient in the air gap, and the ME transducer is placed in the air gap between the magnets. When the harvester is excited, the magnetic circuit moves relative to the ME transducer. The experimental results showed that the generator produced a power of 2.11 mW for an acceleration of 9.8 m s^{-2} at a resonant frequency of 51 Hz.

1.3.5 Comparisons of Transduction Mechanisms

The efficiency of a generator should be simply defined by the standard definition, $\eta = E_{out}/E_{in}$, where E_{out} is the energy delivered to an electrical load and E_{in} is the input energy from the excitation vibrations per cycle. Roundy [48] has proposed a method based upon a standard two-port model of a transducer which enables the different transduction mechanisms to be compared. The analysis uses a coupling coefficient, κ, which is a measure of the efficiency of the conversion from the external vibration energy to the energy stored within the generator and transmission coefficient, λ, which is mathematically identical to the equation for efficiency given above. The transmission coefficient is related to the coupling coefficient and λ_{max} can be found from

$$\lambda_{max} = \frac{\kappa^2}{4 - 2\kappa^2} \tag{1.40}$$

In practice the transmission coefficient depends upon the load resistance which should be chosen to achieve λ_{max}. The maximum power can be found from Eq. (1.41) where ω is the circular frequency of driving vibrations:

$$P_{max} = \lambda_{max}\omega E_{in} \tag{1.41}$$

These coefficients have been derived by Roundy for each of the transduction mechanisms and can be employed to compare them as follows. In the case of the electromagnetic generator, the coupling factor is given in Section 1.3.1. For piezoelectric generators, the following equation applies where d is the piezoelectric strain coefficient (see Section 1.3.2), Y is Young's modulus and ε is the dielectric constant:

$$\kappa^2 = \frac{d^2 Y}{\varepsilon} \tag{1.42}$$

The maximum energy density for both electromagnetic and piezoelectric generators is given by

$$p_{\max} = \frac{\kappa^2 \rho (Q \cdot a)^2}{4\omega} \tag{1.43}$$

where ρ is the density of the proof mass material, Q is the quality factor of the generator and a is the magnitude of acceleration of the excitation vibrations.

The maximum energy density for electrostatic generators is non-linear and depends upon the geometry and operating conditions of the device. Taking the example of an out-of-plane parallel plate capacitor operating in a constant charge mode as described in Section 1.3.3.

$$\kappa^2 = \frac{V_{\text{in}}^2 C_{\max}^2}{V_{\text{in}}^2 C_{\max}^2 + m\omega^2 z^2 C(z)} \tag{1.44}$$

Equation (1.44) gives the coupling coefficient where V_{in} is the input voltage, z the displacement of the top electrode and $C_{\max}$ is the maximum capacitance. It is clear that the capacitance varies with displacement and therefore the coupling coefficient varies throughout the cycle. The average power output density is given by Eq. (1.45) where f is the generator frequency in hertz:

$$p_{\text{ave}} = f \frac{\rho (Q \cdot a)^2}{4\omega^2} \int_{t1}^{t2} \kappa(t) dt \tag{1.45}$$

The coupling coefficient of piezoelectric generators depends mainly on the piezoelectric material used, although the elastic properties of the other materials used in the generator structure may also influence the values. The coupling coefficient of electromagnetic generators is dependant upon the magnetic circuit of the device. In the case of electrostatic generators the coupling coefficient varies with position and device design.

Advantages and disadvantages of each type of transduction mechanism are summarized in Table 1.5.

Since electrostatic and piezoelectric transducers are compatible with MEMS, they are more suitable to be deployed in micro- or nano-scale systems while electromagnetic and magnetostrictive transducers are suitable for macro-scale systems. Roundy et al. [21] calculated the theoretical maximum energy density of the first three transducers. It was concluded that piezoelectric and electromagnetic transducers have similar energy density which is about 10 times that of electrostatic transducers.

Table 1.6 lists some commercially available vibration-based micro-generators. To the date, only generators with electromagnetic and piezoelectric transducers can

Table 1.5 Comparisons of different transduction mechanisms of kinetic energy harvesters

Type	Advantages	Disadvantages
Electromagnetic	• No external voltage source • No mechanical constraints needed • High output current	• Difficult to integrate with MEMS fabrication process • Poor performance in micro-scale • Low output voltage
Piezoelectric	• Simple structure • No external voltage source • Compatible with MEMS • High output voltage • No mechanical constraints needed	• Thin films have poor coupling • Poor mechanical properties • High output impedance • Charge leakage • Low output current
Electrostatic	• Easy to integrate with MEMS fabrication process • High output voltage	• Mechanical constraints needed • External voltage source or pre-charged electret needed • High output impedance • Low output current
Magnetostrictive	• Ultra-high coupling coefficient • High flexibility	• Non-linear effect • May need bias magnets • Difficult to integrate with MEMS fabrication process

Table 1.6 Summary of kinetic energy harvesters available on the market

Model	*f* (Hz)	Excitation level (m s^{-2})	Total mass (g)	Volume (mm^3)	*P* (mW)	Transducer
		Mide Technology Corporation [49]				
Volture PEH20w	80–175[a]	13.7	85.14	39,963[b] 388.55[c]	2–24	PZ
Volture PEH25w	50–140[a]	13.7	85.14	40,543[b] 194.27[c]	2.5–24	PZ
		Perpetuum Ltd. [50]				
PMG-17	100/120	9.8	655	522,682[b]	45	EM
PMG-27	17.2	0.49	400	467,711[b]	4	EM

PZ Piezoelectric, *EM* Electromagnetic

[a] Tunable by changing the length of the cantilever manually

[b] Total device

[c] Piezo wafer

be found on the market, which indicates that these two transducers are more feasible in practice.

1.4 Introduction to Adaptive Kinetic Energy Harvesting

Mathematical analysis presented in Section 1.2 showed that maximum power is generated when the resonant frequency of the generator matches the frequency of the ambient vibration. The limitation to this feature is that the generator is, by definition, designed to work at a single frequency. A high Q-resonance means very limited practical bandwidths over which energy can be harvested. If the resonant frequency of the generator does not match the ambient vibration frequency, the generated power drops dramatically. Most reported kinetic energy harvesters are designed to work only at one particular frequency [2]. For applications such as moving vehicles, human movement and wind-induced vibration where the frequency of ambient vibration changes occasionally, the efficiency of generators with one fixed resonant frequency is significantly reduced since the generator will not always be at resonance. This drawback must be overcome if kinetic energy harvesters are to be widely applicable in powering wireless systems. Therefore, adaptive energy harvesting is developed to increase operational frequency range of kinetic energy harvesters. To date, there are, in general, two approaches to achieving adaptive energy harvesters.

The first approach is to adjust, or tune, the resonant frequency of a single generator so that it matches the frequency of the ambient vibration at all times. This can be achieved by changing the mechanical characteristics of the structure or electrical load of the generator. Resonant frequency tuning methods can be classified as intermittent and continuous tuning [3]. Intermittent tuning is defined as a tuning mechanism that operates periodically. This approach only consumes power during the tuning operation and uses negligible energy once the generator is matched to the frequency of the ambient vibrations. Continuous tuning is defined as a tuning mechanism that is continuously applied even if the resonant frequency equals the ambient vibration frequency. The second approach is to widen the bandwidth of the generator. This can be achieved by, for example, employing

- an array of structures each with a different resonant frequency;
- an amplitude limiter;
- coupled oscillators;
- non-linear (e.g. magnetic) springs;
- bi-stable structures;
- a large inertial mass (large device size) with a high degree of damping.

Details of these two types of adaptive kinetic energy harvesters will be studied and compared in the following three sections.

1.5 Strategies to Tune Resonant Frequency

1.5.1 Evaluating Tuning Approaches

Selection of tuning approaches will depend upon the application but in general some key factors for evaluating a tuning mechanism for adjusting the resonant frequency of kinetic energy harvesters are as follows:

- The energy consumed by the tuning mechanism should be as small as possible and must not exceed the energy produced by the generator.
- The mechanism should achieve a sufficient operational frequency range.
- The tuning mechanism should achieve a suitable degree of frequency resolution.
- The generator should have as high as possible Q-factor to achieve maximum power output and the strategy applied should not increase the damping, i.e. decrease Q-factor, over the entire operational frequency range.

Resonant frequency can be tuned by both mechanical and electrical methods. Mechanical tuning alters the resonant frequency by changing mechanical properties of the structure. Electrical tuning alters the resonant frequency by adjusting the electrical load. The principles of both methods and existing approaches to realize them are described in the following sections.

1.5.2 Mechanical Tuning Methods

As most reported vibration energy harvesting devices are based on a cantilever [2], focus will be on this structure in following theoretical analyses of mechanical tuning. Principles demonstrated are, however, generally applicable to all mechanical resonator structures. Tuning mechanisms covered in this section are

- changing dimensions;
- moving the centre of gravity of proof mass;
- variable spring stiffness;
- straining the structure.

After a brief analysis of the theory, a comprehensive review of each mechanical tuning mechanism reported in the literature to date is presented.

The resonant frequency of a spring-mass structure is given by

$$f_{\mathrm{r}} = \frac{1}{2\pi}\sqrt{\frac{k}{m}} \tag{1.46}$$

where k is the spring constant and m is the inertial mass. When tuning the resonant frequency of the generator, either the spring constant or the mass can be varied.

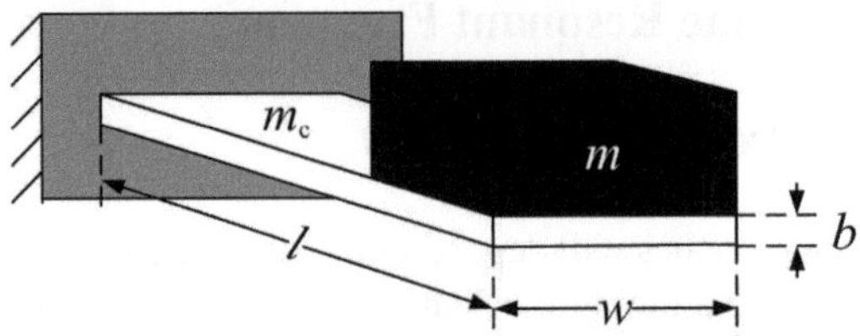

Fig. 1.11 Cantilever with a tip mass

The spring constant of a resonator depends on its materials and dimensions. For a cantilever with a mass at the free end (Fig. 1.11), the resonant frequency is given by [51]

$$f_r = \frac{1}{2\pi}\sqrt{\frac{Ywh^3}{4l^3(m+0.24m_c)}} \tag{1.47}$$

where Y is Young's modulus of the cantilever material, w, h and l are the width, thickness and length of the cantilever, respectively, and m_c is the mass of the cantilever. The resonant frequency can be tuned by adjusting any of these parameters. In addition, it is important to mention that if actuators are involved in changing the mechanical properties of the resonant structure, the tuning mechanisms can be operated by a control system to automatically tune the generator. However, the energy cost of the actuator must be considered.

1.5.2.1 Changing Dimensions

It is difficult to change the width, w, and thickness, h, of a cantilever after it is made while changing its effective length, l, is feasible. The approach requires that the cantilever base clamp be released and re-clamped in a new location along the length of the beam thereby changing the effective length (and hence frequency). There is no power required to maintain the new resonant frequency. It is a method of intermittent tuning. Furthermore, as the resonant frequency is inversely proportional to $l^{3/2}$, modifying l can significantly change f_r.

Suppose l is the original length of the cantilever and l' is the modified length of the cantilever, $l' = l + \Delta$, where Δ is the difference between them. The mass of the cantilever is then changed to $m_c' = whl'\rho$, where ρ is the density of the cantilever material while the original mass of cantilever is $m_c = whl\rho$. Then, the new resonant frequency becomes

$$f_r' = \frac{1}{2\pi}\sqrt{\frac{Ywh^3}{4l'^3(m+0.24m_c')}} = \frac{1}{2\pi}\sqrt{\frac{Ywh^3}{4(l+\Delta)^3\{m+0.24[wh(l+\Delta)]\rho\}}} \tag{1.48}$$

And the ratio of the tuned frequency to the original frequency called the normalized resonant frequency is

$$\frac{f_{\rm r}'}{f_{\rm r}} = \sqrt{\frac{l^3(m + 0.24m_{\rm c})}{l'^3(m + 0.24m_{\rm c}')}} = \sqrt{\frac{l^3(m + 0.24whl\rho)}{(l + \Delta)^3\{m + 0.24[wh(l + \Delta)]\rho\}}} \tag{1.49}$$

Figure 1.12 shows the normalized resonant frequency with the variation of cantilever length where a negative Δ/l means the new cantilever beam is shorter than its original length and thus has a higher resonant frequency. A positive Δ/l means the cantilever beam has been lengthened giving a lower resonant frequency. Figure 1.12 shows it is more efficient to tune the resonant frequency by shortening the cantilever beam.

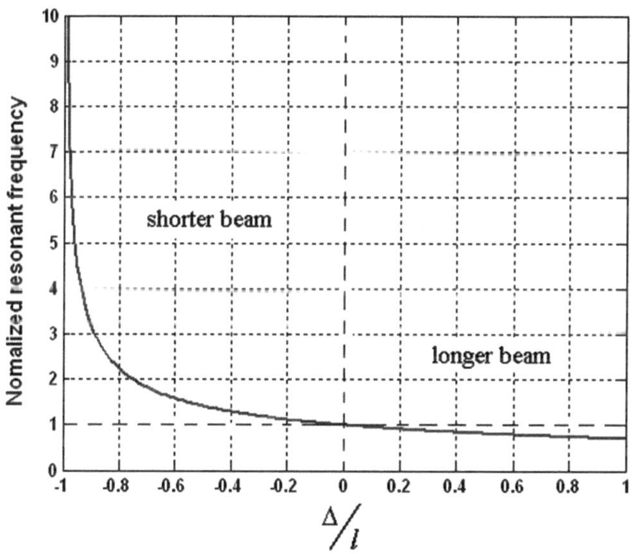

Fig. 1.12 Normalized resonant frequency with variation of cantilever lengths

An example of this approach is described in a patent by Gieras et al. [52]. Figure 1.13 shows the side view of the proposed device. The electromagnetic generator consists of a cantilever with a set of magnets fixed to its free end. The cantilever is clamped to a base using screws. A coil is placed between the magnets to pick up output power. A slider is connected to a linear actuator which moves the slider back and forth to adjust the effective length of the cantilever, L, and hence the resonant frequency of the generator.

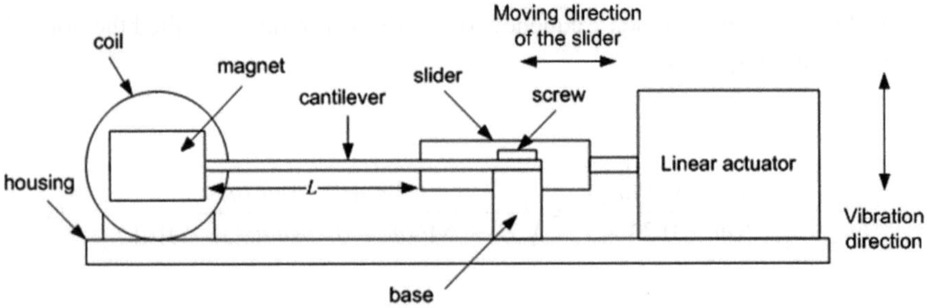

Fig. 1.13 Side view of a self-adjustable energy-harvesting system with variable effective lengths

1.5.2.2 Moving Centre of Gravity of Proof Mass

Once a generator has been fabricated, it is difficult to subsequently add or remove mass. However, the resonant frequency of a cantilever structure can be adjusted by moving the centre of gravity of the inertial mass. Figure 1.14 shows the side view of a cantilever with a mass on the free end.

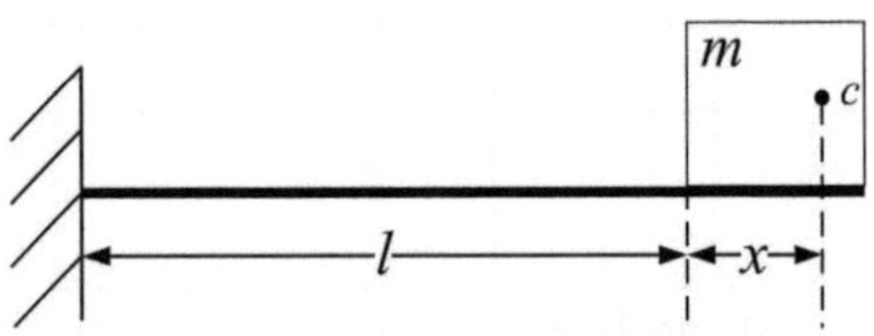

Fig. 1.14 Side view of a cantilever-mass structure

The length of the cantilever without the mass is l and the proof mass on its free end is m. The centre of gravity of the proof mass is located at c and the distance between c and the end of the cantilever is x. The tuned resonant frequency of this structure can be approximated as [53] (Fig. 1.15)

$$f_r' = \frac{1}{2\pi}\sqrt{\frac{Ywh^3}{12ml^3}\cdot\frac{r^2+6r+2}{8r^4+14r^3+\frac{21}{2}r^2+4r+\frac{2}{3}}} \tag{1.50}$$

where w and h are the width and thickness of the cantilever, respectively, and $r = x/l$.

The resonant frequency of a cantilever-based generator, considering that the mass of the cantilever beam is negligible compared to the proof mass, Eq. (1.47), can be rewritten as

$$f_r = \frac{1}{2\pi}\sqrt{\frac{Ywh^3}{4l^3m}} \tag{1.51}$$

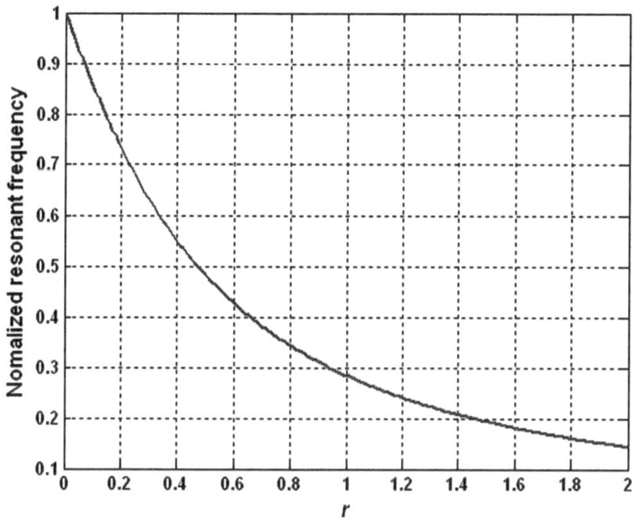

Fig. 1.15 Resonant frequency with variation of centre of gravity positions

Hence, the ratio of the tuned frequency to the original frequency is

$$\frac{f_r'}{f_r} = \sqrt{\frac{1}{3} \cdot \frac{r^2 + 6r + 2}{8r^4 + 14r^3 + \frac{21}{2}r^2 + 4r + \frac{2}{3}}} \tag{1.52}$$

Figure 1.14 shows the normalized resonant frequency with variation of the position of the centre of gravity of the proof mass. The closer the centre of gravity of the proof mass is from the end of the cantilever, the higher the resonant frequency.

Wu et al. [54] reported a piezoelectric generator using this principle as shown in Fig. 1.16. The proof mass of this device consists of a fixed mass and a movable mass. The position of the centre of gravity of the proof mass could be adjusted by changing the position of the movable mass. A fastening stud was used to fix the movable mass when tuning was finished. The size of the fixed mass is 10 mm × 12 mm × 38 mm and the movable mass is an M6 screw of length of 30 mm. The resonant frequency of the device was tuned from 180 to 130 Hz by moving the screw from one end to the other end (Fig. 1.17). The output voltage dropped with increasing resonant frequency. This approach is suitable for fine frequency tuning of the generator before installation and the vibration frequency in the working environment is not time varying. If the vibration frequency changes during operation, an actuator has to be employed on the cantilever to change the position of the movable mass, which increases the complexity of the generator.

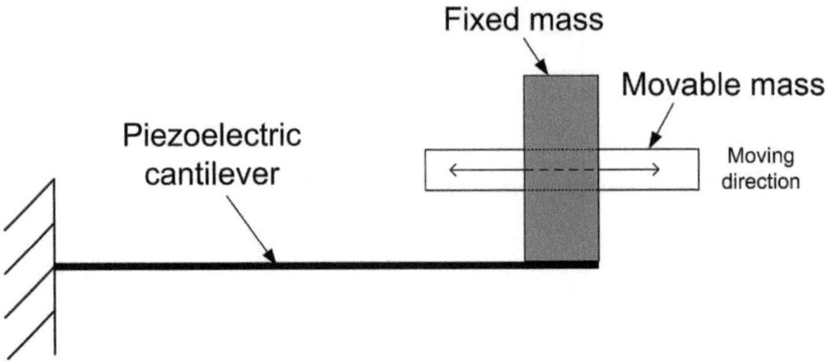

Fig. 1.16 A piezoelectric cantilever with movable mass

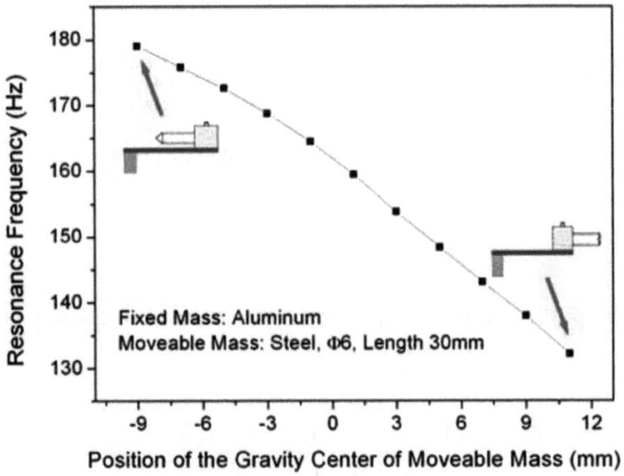

Fig. 1.17 Experimental result of frequency adjustment [54]

1.5.2.3 Variable Effective Spring Stiffness

Another commonly used method to tune the resonant frequency is to soften the spring stiffness. The principle is to apply a "negative" spring in parallel to the mechanical spring. Therefore, the effective spring constant of such device, k_{eff}, becomes

$$k_{\mathrm{eff}} = k_{\mathrm{m}} + k_{\mathrm{a}} \tag{1.53}$$

where k_{m} is the mechanical spring constant and k_{a} is an additional "negative" spring stiffness as shown in Fig. 1.18. The modified frequency becomes

$$f_{\mathrm{r}} = \frac{1}{2\pi}\sqrt{\frac{k_{\mathrm{eff}}}{m}} = \frac{1}{2\pi}\sqrt{\frac{k_{\mathrm{m}} + k_{\mathrm{a}}}{m}} \tag{1.54}$$

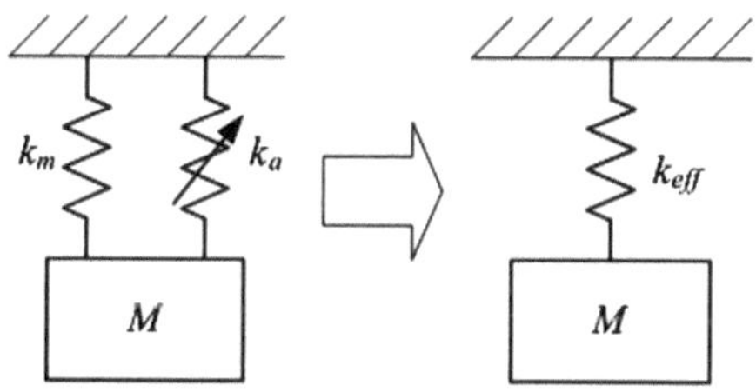

Fig. 1.18 Model of devices with softened spring stiffness

The negative spring k_a can be applied electrostatically, piezoelectrically, magnetically or thermally. Examples of these approaches are described below. Most of these examples are tunable resonators and not energy harvesters but the principles are identical. It is important to note, however, that the additional inertial mass present in an energy harvester (as opposed to the purely resonant structures) will reduce the tuning effectiveness and increase the power required to tune compared to the values quoted. It should also be noted that the following variable spring stiffness devices are all continuously operated except the one on which the negative spring is applied magnetically.

Electrostatic Methods

Scheibner et al. [55, 56] reported a vibration detector consisting of an array of eight comb resonators each with a different base resonant frequency. A single resonator is shown in Fig. 1.19. Each resonator comb is tuned by electrostatically softening the structure by applying a tuning voltage to the electrodes marked "V_{Tun}". The device was designed so that the resonator array had overlapping tuning ranges which allowed continuous measurements in the frequency range of the device from 1 to 10 kHz. Figure 1.20 shows the tuning range of each resonator. The tuning voltage varied from 0 to 35 V. The total size of the sensor chip is 7 mm × 10 mm.

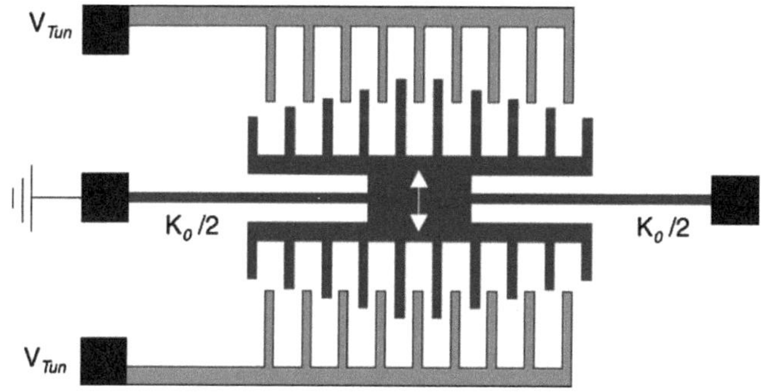

Fig. 1.19 Resonance tuning by electrostatic softening [56]

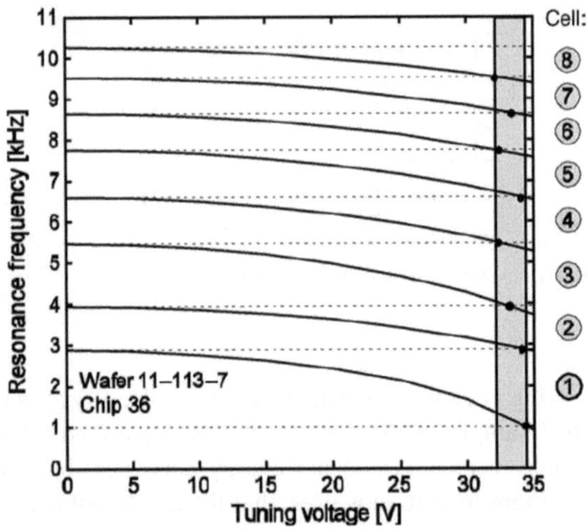

Fig. 1.20 Experimental results of the resonance tuning of the array [56]

Adams et al. [57] realized a tuning range from 7.7 to 146% of the central frequency of 25 kHz of a resonator with a single comb structure (Fig. 1.21). Figure 1.22 shows the tuning ranges of two of their devices under the driving voltage between 0 and 50 V. The total size was not mentioned in the chapter but is estimated from the SEM scale to be no larger than 500 μm × 500 μm.

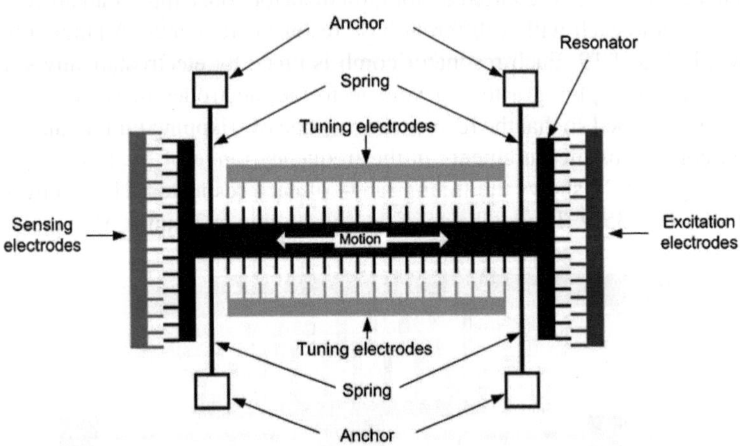

Fig. 1.21 Schematic diagram of single comb structure (after [57])

Lee et al. [58] presented a frequency-tunable comb resonator with curved comb fingers (Fig. 1.23). Fingers of the tuning comb were designed to be curved shape to generate a constant electrostatic stiffness or linear electrostatic force that is independent of the displacement of the resonator under a control voltage. Experimentally, the resonant frequency of a laterally driven comb resonator with 186 pairs of

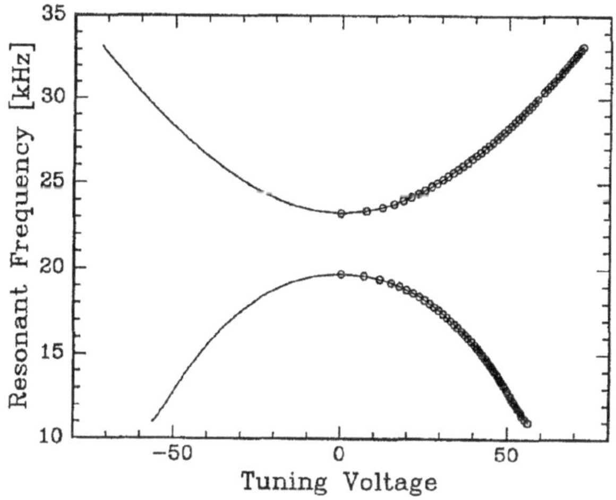

Fig. 1.22 Experimental results of resonance tuning of single comb structure [57]

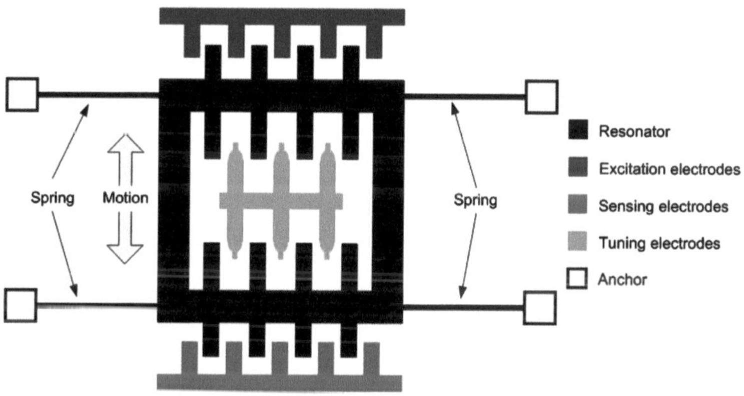

Fig. 1.23 Schematic diagram of comb resonator with curved tuning fingers (after [58])

curved contour fingers was reduced by 55% from the initial frequency of 19 kHz under a bias voltage of 150 V (Fig. 1.24). The corresponding effective stiffness was decreased by 80% from the initial value of 2.64 N/m. The total size of the resonator is 460 μm × 840 μm. It was concluded that the closed-form approach of the comb-finger profile can be applied to other comb-shaped actuators for frequency control while achieving linear electrostatic stiffness with respect to displacement.

Piazza et al. [59] developed a micromachined, piezoelectrically actuated and sensed, high-Q single-crystal silicon (SCS) resonator with voltage-tunable centre frequency (Fig. 1.25). Piezoelectric transduction was integrated with capacitive fine-tuning of the resonator centre frequency to compensate for any process variations. The resonant frequency could be tuned by 6 kHz based on an untuned resonant frequency of 719 kHz by applying an electrostatic force beneath the cantilever

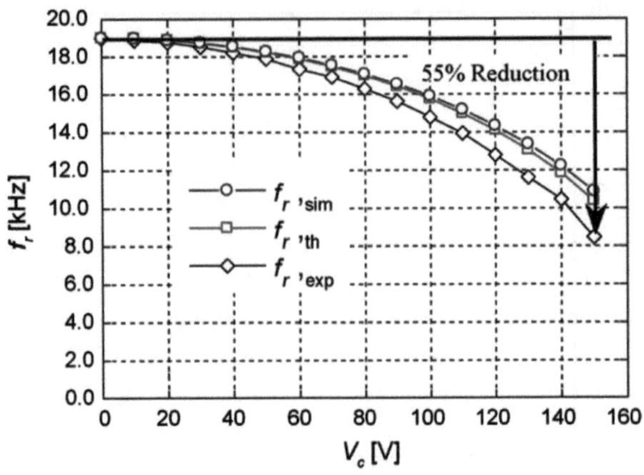

Fig. 1.24 Experimental results of resonance tuning of comb resonator with curved tuning fingers [58]

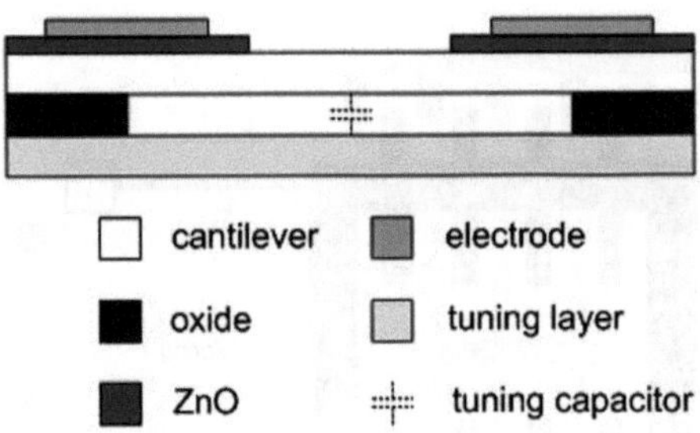

Fig. 1.25 Voltage-tunable, piezoelectrically transduced SCS resonators: Q-enhanced configuration (after [59])

(Fig. 1.26). The driving voltage varied from 0 to 20 V. The dimensions of this resonator are 200 μm × 20 μm × 4.2 μm.

Yao and MacDonald [60] compared frequency tuning by applying either axial force or transverse force on the resonator electrostatically as shown in Fig. 1.27. Frequency tuning by applying transverse force was tested experimentally. It was found that the resonant frequency may increase or decrease with the applied tuning voltage depending on where the tuning electrode is placed with respect to the excitation electrode and the resonating rod. When the tuning electrode was placed on the same side of the excitation electrode as indicated in Fig. 1.27b, the resonant frequency decreased with the increase of applied voltage. When the tuning electrode was placed on the opposite side of the excitation electrode as indicated in Fig. 1.27c, the resonant frequency increased with the increase of applied voltage.

A micromachined resonator having an out-of-plane natural resonant frequency of 149.5 kHz was tuned to 139.5 kHz by applying a DC tuning voltage of 30 V as shown in Fig. 1.28. The actual dimensions of these devices were not mentioned. Similar idea was later patented by Thiesen and O'Brian [61].

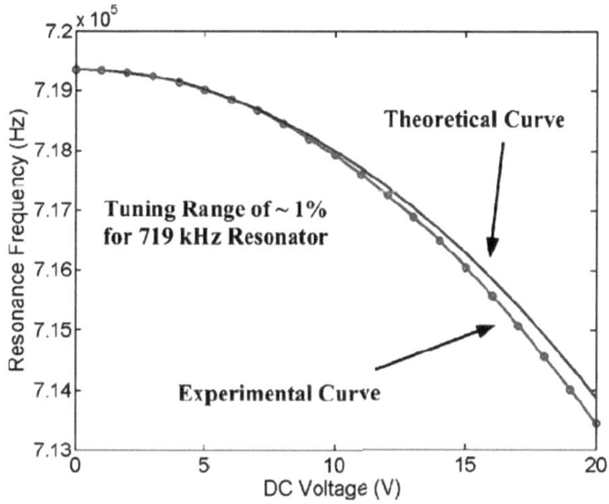

Fig. 1.26 Experimental results of resonance tuning of electrostatic fine-tuning characteristic for a 719 kHz piezo-resonator [59]

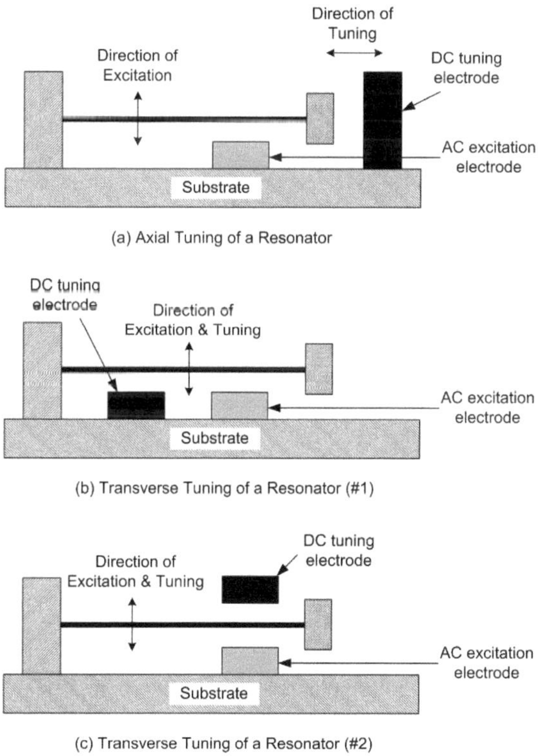

Fig. 1.27 Schematic drawing of a simple resonator showing axial loading (**a**), and transverse loading with the excitation and the tuning electrode on the same side (**b**) and on the opposite side (**c**) of the resonating rod [60]

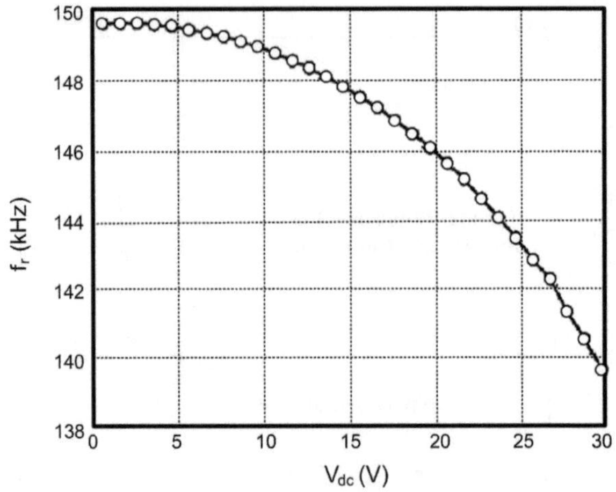

Fig. 1.28 Measured resonant frequency versus the tuning DC voltage with an untuned resonant frequency of 149.5 kHz (tuning mechanism as in Fig. 1.27b) [60]

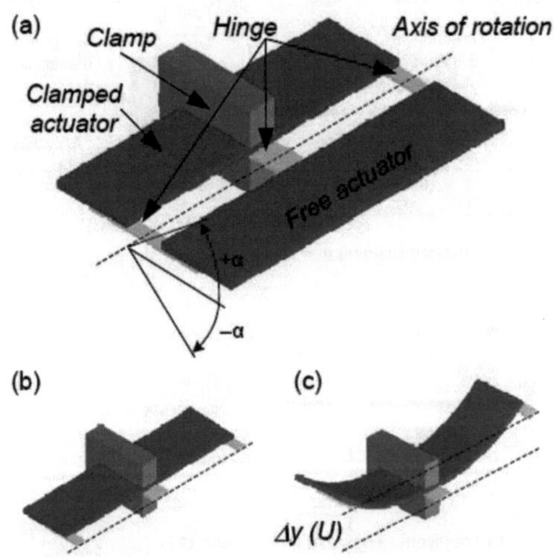

Fig. 1.29 (**a**) Schematic of the resonator (**b**) cross-section without applied voltage and (**c**) with applied voltage [62]

Piezoelectric Methods

Peters et al. [62] reported a tunable resonator, shown in Fig. 1.29a, potentially suitable as a resonator structure for vibration energy harvesting. The resonant frequency is tuned by mechanically stiffening the structure using piezoelectric actuators. A piezoelectric actuator was used because piezoelectric materials can generate large forces with low power consumption. Two actuators, one clamped and one free, are connected together. The free actuator can oscillate around the axis of rotation if a suitable excitation is applied to the clamp. The stiffness of the structure was increased by applying an electrical potential to both actuators which changes the shape of the structure as shown in Fig. 1.29c. Thus, the natural frequency of the rotational mass-spring system increased. The tuning voltage was chosen to be ±5 V leading to a measured resonance shift of ±15% around the initial resonant frequency of 78 Hz, i.e. the tuning range was from 66 to 89 Hz (Fig. 1.30).

Wischke et al. [63] presented an electromagnetic vibration scavenger that exhibits a tunable resonant frequency as shown in Fig. 1.31. The resonant frequency can be tuned by applying a static electrical field on the piezoelectric cantilever. This feature is originated from exploiting the elastostriction of the utilized piezoelectric bimorph suspension. It is demonstrated that the resonant frequency has been tuned from 267 to 323 Hz when a tuning voltage between −100 and 260 V is applied.

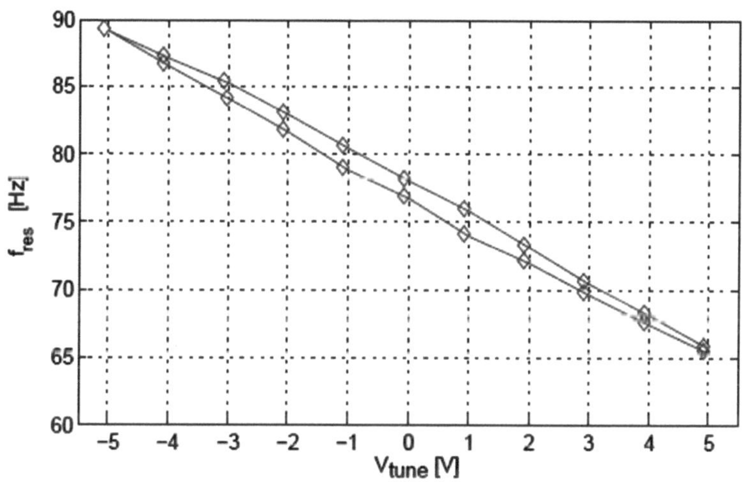

Fig. 1.30 Measured resonant frequency versus applied tuning voltage [62]

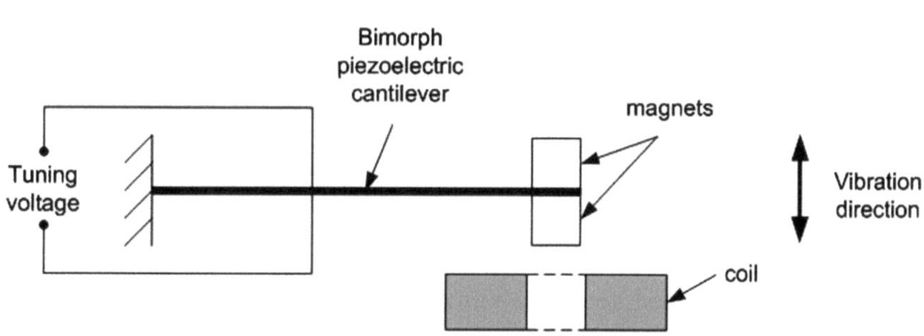

Fig. 1.31 A piezoelectrically tunable electromagnetic generator (after [63])

Magnetic Methods

Challa et al. [64] reported an intermittently tuned piezoelectric micro-generator (Fig. 1.32), 50 cm^3 in volume, with a frequency range of 22–32 Hz based on an original resonant frequency of 26 Hz. The tuning was realized by applying a magnetic force perpendicular to the cantilever generator as shown in Fig. 1.33. The resonant frequency of the generator can be tuned by changing the distance between the two sets of tuning magnets and the fixed magnets. The maximum tuning distance was 3 cm. Experimentally, the generator produced 240–280 μW power at 0.8 m s^{-2} acceleration. However, the tuning mechanism had the unwanted side effect of varying damping over the frequency range as shown in Fig. 1.33. The device was made of discrete components. The dimension of the piezoelectric cantilever is 34 mm × 20 mm × 0.92 mm and the effective mass is 45.8 g.

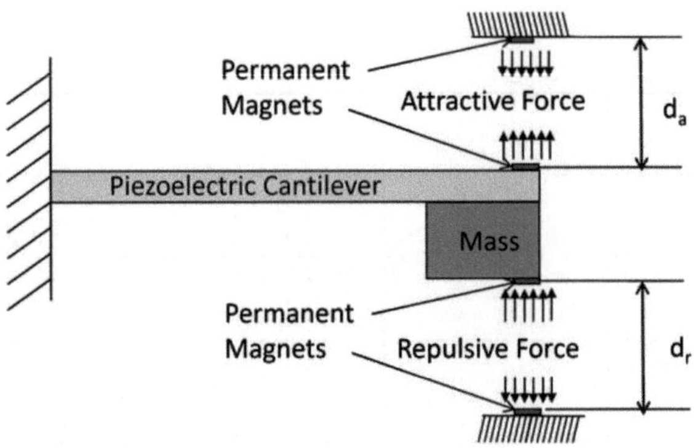

Fig. 1.32 Schematic of the tunable piezoelectric generator [64]

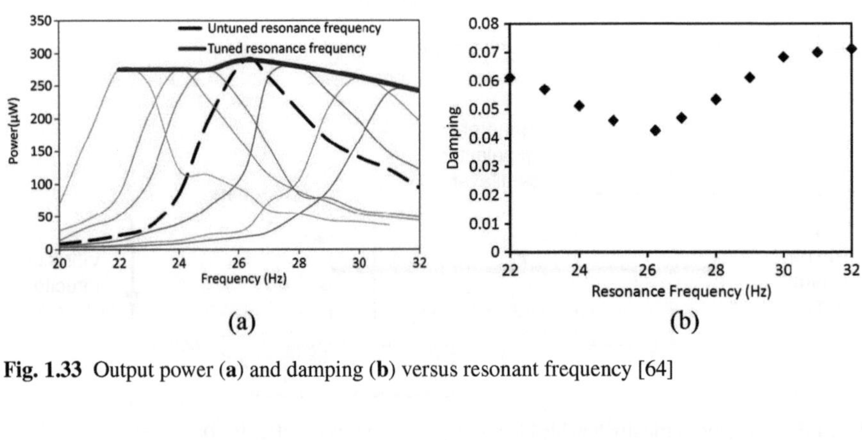

Fig. 1.33 Output power (**a**) and damping (**b**) versus resonant frequency [64]

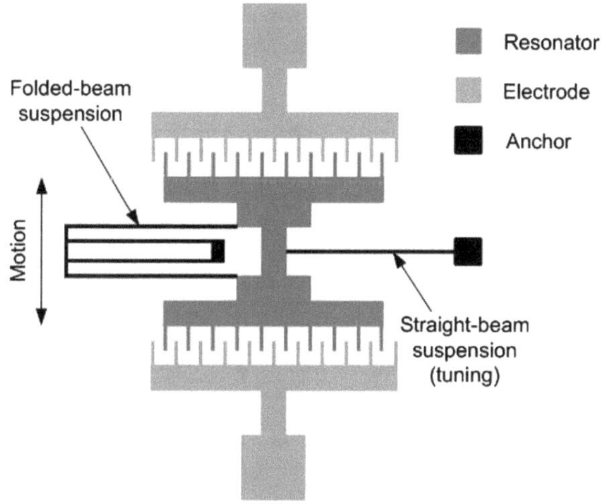

Fig. 1.34 Schematic diagram of a comb-shaped micro-resonator with a straight-beam for active frequency tuning via localized stressing effects (after [65])

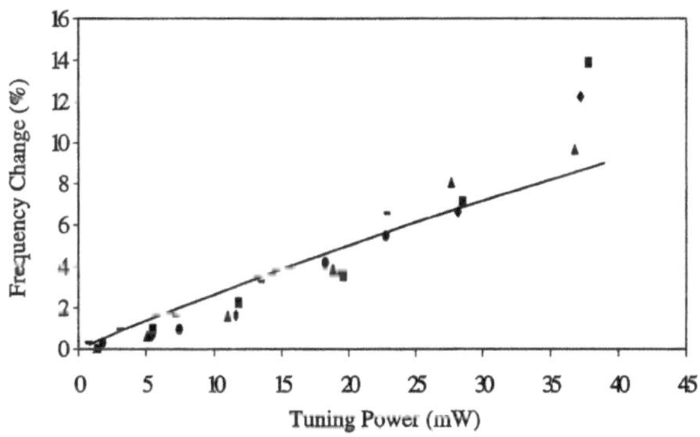

Fig. 1.35 Measured frequency change versus tuning power [65]

Thermal Methods

Remtema and Lin [65] applied a thermal stress on a straight-beam spring using a resistive heater (Fig. 1.34), which resulted in a maximum 6.5% frequency change based on a resonant frequency of 31 kHz with a maximum temperature at 255°C. The power consumption during the process was 25 mW. Figure 1.35 shows the percentage change of resonant frequency with variation of power consumed in tuning. The size of the device is estimated to be less than 500 μm × 700 μm from the authors' description.

Syms [66] reported frequency tuning by applying constrained thermal expansion on a simple unfolded resonator (Fig. 1.36). The tuning range was from −25 to +50% with power consumption from 1.5 to 10 mW (Fig. 1.37). The tuning sensitivity obtained with this tuning method was 33% mW^{-1}. It is estimated from the annotation in Fig. 1.36 that the device is no larger than 3000 μm × 3000 μm.

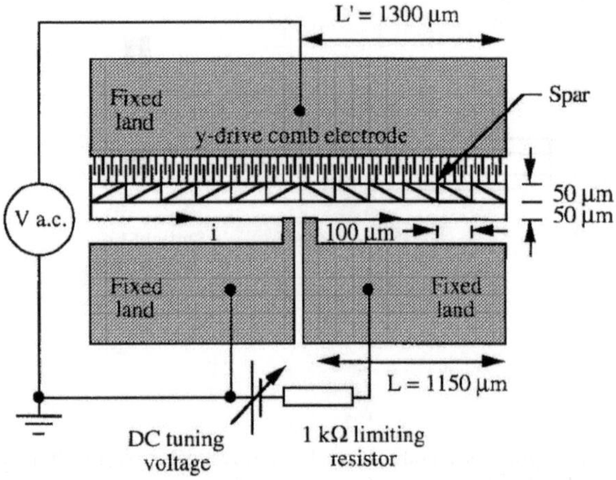

Fig. 1.36 Layout and connection of laterally resonant comb-drive actuator used for tuning experiments [66]

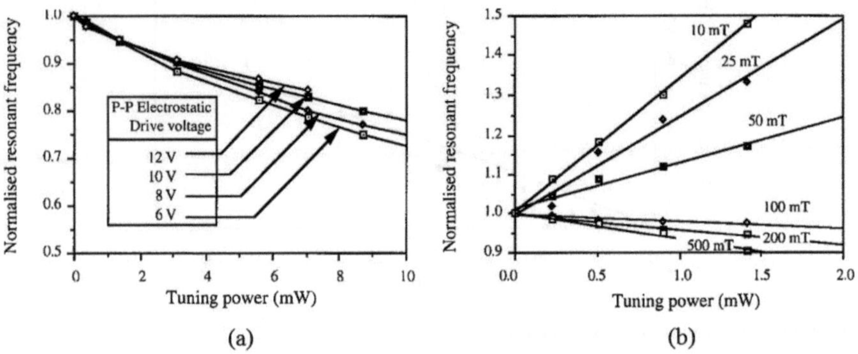

Fig. 1.37 Variation of resonant frequency with tuning power (**a**) at different electrostatic drive voltages and (**b**) gas pressures [66]

The thermal approach is unlikely to be practical for energy harvesting since it consumes too much power compared to the power generated by a kinetic energy harvester and is a continuous tuning mechanism.

1.5.2.4 Straining the Structure

The effective stiffness of the structure can be varied by applying a stress and therefore placing it under strain. The following theoretical analyses focus on straining a cantilever and a double-clamped beam. The resonant frequency of both these structures can be tuned by applying an axial load. In vibration energy harvesting, most devices are based on beam structures, especially the cantilever (Fig. 1.38). An axial tensile load applied to a beam (Figs. 1.38a and 1.39a) increases the resonant frequency while an axial compressive load (Figs. 1.38b and 1.39b) decreases the resonant frequency of the beam.

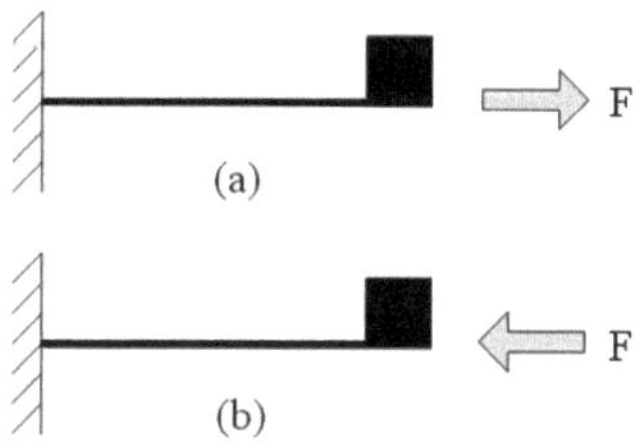

Fig. 1.38 Axial tensile (**a**) and compressive (**b**) loads on a cantilever

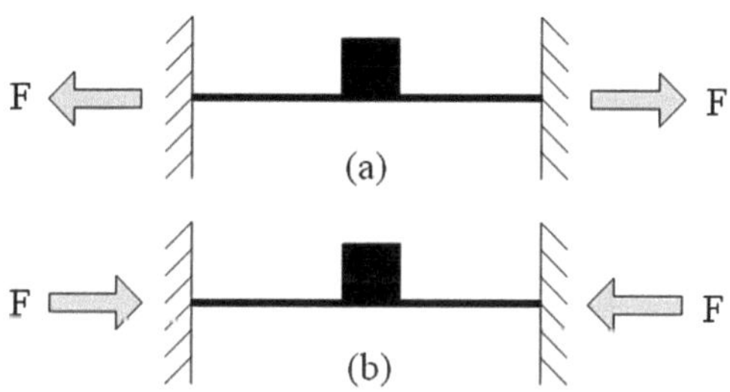

Fig. 1.39 Axial tensile (**a**) and compressive (**b**) loads on a double-clamped beam

An approximate formula for the resonant frequency of a uniform cantilever in mode i with an axial load, f_{ri}, is given by [51]

$$f'_{ri} = f_{ri} \cdot \sqrt{1 + \frac{F}{F_b} \cdot \frac{\lambda_1^2}{\lambda_i^2}} \tag{1.55}$$

where f_r is the resonant frequency in mode i without load, F is the axial load and F_b is the axial load required to buckle the beam, i.e. to cause the fundamental resonant frequency zero. F is positive for a tensile load and negative in the compressive case. Variable λ_i is a dimensionless load parameter which is a function of the beam boundary conditions applied to the cantilever for the ith mode of the beam. It is given by the ith positive solution of Eq. (1.56) for a cantilever and of Eq. (1.57) for a double-clamped beam [51].

$$\cos\lambda \cdot \cosh\lambda + 1 = 0 \tag{1.56}$$

$$\cos\lambda \cdot \cosh\lambda - 1 = 0 \tag{1.57}$$

The majority of cantilever-based energy harvesters operate in the fundamental flexural mode (mode 1); the resonant frequency of a uniform cantilever in mode 1 with an axial load, f_{r1}, is given by

$$f'_{r1} = f_{r1} \cdot \sqrt{1 + \frac{F}{F_b}} \tag{1.58}$$

The ratio of the tuned frequency to the original frequency is

$$\frac{f'_{r1}}{f_{r1}} = \sqrt{1 + \frac{F}{F_b}} \tag{1.59}$$

The buckling load F_b of a cantilever and a clamped–clamped beam are given by Eqs. (1.60) and (1.61), respectively [60]:

$$F_{b\_can} = \frac{\pi^2 Y w h^3}{48 l^2} \tag{1.60}$$

$$F_{b\_dcb} = \frac{\pi^2 Y w h^3}{3 l^2} \tag{1.61}$$

where Y is Young's modulus of the material of the cantilever and w, h and l are the width, thickness and length of the cantilever, respectively.

Figure 1.40 shows the change in resonant frequency of a cantilever with an applied axial load. It shows that a compressive load is more efficient in frequency tuning than a tensile load. If the compressive force is larger than the buckling load, the cantilever beam will buckle and no longer oscillate in mode 1. If a very large

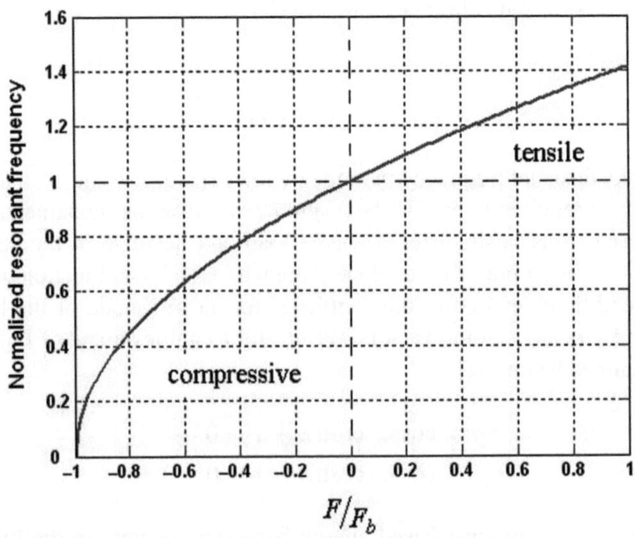

Fig. 1.40 Normalized resonant frequency with variation of axial loads

tensile force is axially applied to the cantilever, i.e. much greater than the buckling load, the resonant frequency will approach that of a straight-tensioned cable as the force associated with the tension in the cantilever becomes much greater than the beam stiffness.

Double-Clamped Beam Structures

Cabuz et al. [67] realized resonant frequency tuning by an applying axial force on a micromachined resonant beam electrostatically as shown in Fig. 1.41. One end of the resonator was clamped on a fixed support while the other end was connected to a movable support. The movable support could rotate around a torsion bar as a voltage was applied across two tuning electrodes. The torsion bar converted the vertical tuning motion into an axial force along the resonator. Upward rotation induces a compressive stress in the resonator while downward rotation induces a tensile stress. The tuning range was 16 Hz based on a centre frequency of 518 Hz (Fig. 1.42) with driving voltage from 0 to 16 V. The dimensions of the resonator are 1000 μm $\times$ 200 μm $\times$ 3 μm and the dimensions of the movable support are 12.5 mm^2 $\times$ 0.3 mm. This is an example of continuous tuning.

Leland and Wright [68] successfully tuned the resonant frequency of a vibration-based piezoelectric generator by applying an axial compressive preload directly on the cantilever (Fig. 1.43). The tuning range was from 200 to 250 Hz. Experimentally, the generator produced an output power between 300 and 400 μW at an acceleration of 9.8 m s^{-2}. It was determined that a compressive axial preload could reduce the resonance frequency of a vibration energy scavenger by up to 24% but it also increased the total damping (Fig. 1.44). The piezoelectric bimorph has

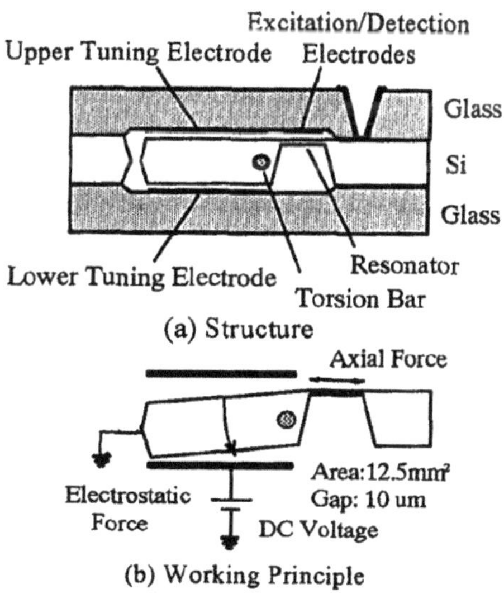

Fig. 1.41 Structure for fine resonance frequency tuning at device level by an electrostatically induced axial force [67]

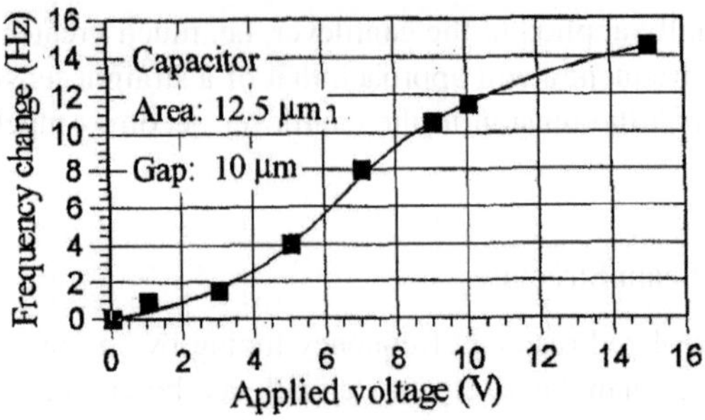

Fig. 1.42 Resonant frequency change versus applied voltage [67]

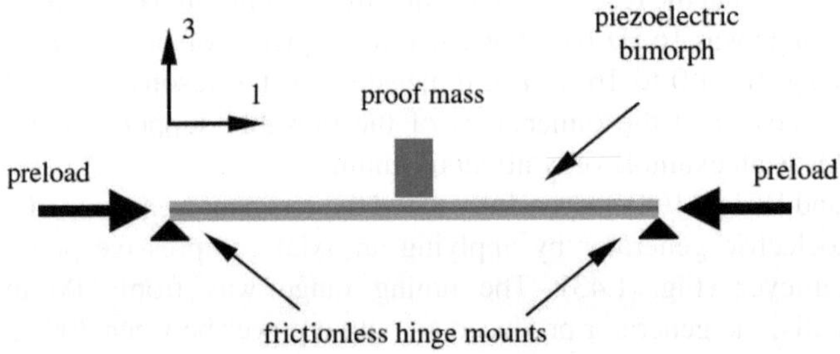

Fig. 1.43 Schematic of a simply supported piezoelectric bimorph vibration energy scavenger [68]

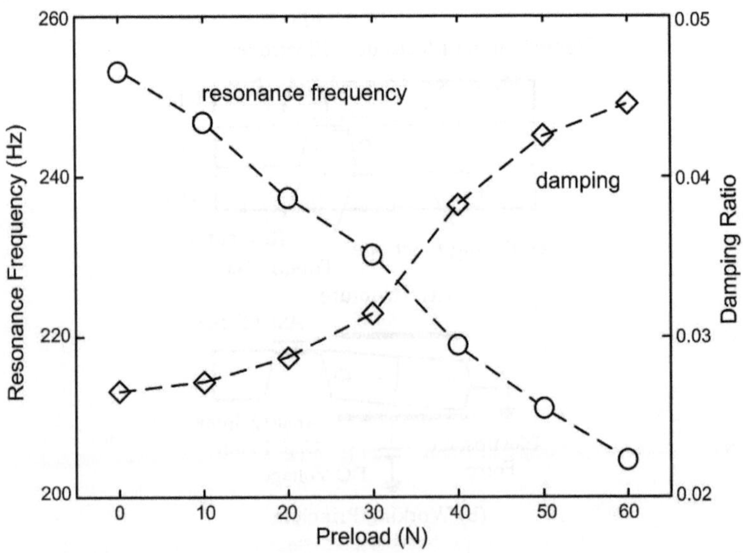

Fig. 1.44 Resonance frequency and damping versus preload [68]

dimensions of 31.7 mm × 12.7 mm × 0.509 mm and the weight of the proof mass is 7.1 g. This is an example of intermittent tuning, but it is not automated and has to be done manually.

Cantilever Structures

Mukherjee [69] patented the idea of applying axial force to a vibrating cantilever beam sensing element using electrostatic force. The resonator consisted of two sets of comb-like structures (Fig. 1.45). The set closer to the anchor was used for sensing while the other set was used for frequency tuning. A voltage was applied between the two fixed tuning electrodes and the structure at the free end to apply an axial tensile or compressive end load to the cantilever. The resonant frequency of the beam was approximately 15.5 kHz. This is an example of continuous tuning which achieved a tuning range of −0.6 to 3.3% of its untuned resonant frequency, i.e. about 600 Hz. The cantilever buckled when 50 V_{DC} was applied to provide a compressive force. This is an example of continuous tuning.

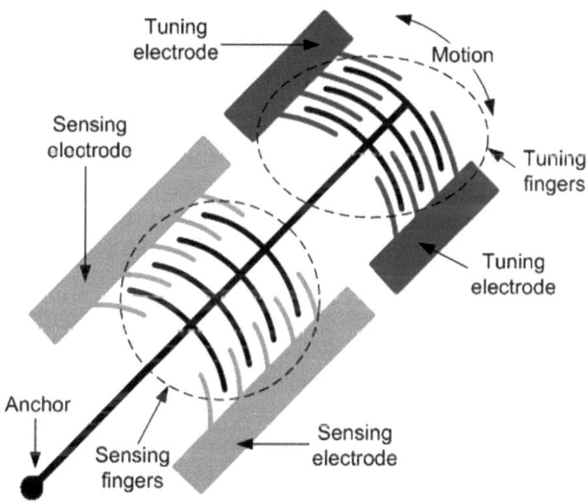

Fig. 1.45 Resonator with actuator at the free end

Hu et al. [70] theoretically investigated an axial preloading technique to adjust the behaviour of a piezoelectric bimorph. It was found analytically that this mechanism can improve the performance of the piezoelectric bimorph at varying frequency vibrations. A method for applying an axial preload to a piezoelectric bimorph was suggested and is shown in Fig. 1.46. It comprises a mechanical bolt running through the central metal layer and fixed at the left-hand side edge wall. A capped stiff metal plate was attached to the bolt at the free end of the cantilever. A clockwise torsion of the bolt can produce a compressive preload to the bimorph and, conversely, an anticlockwise torsion of the bolt produces a force to pull the capping plate towards the right-hand side, which can generate a tensile preload to the bimorph. This is an example of manual intermittent tuning.

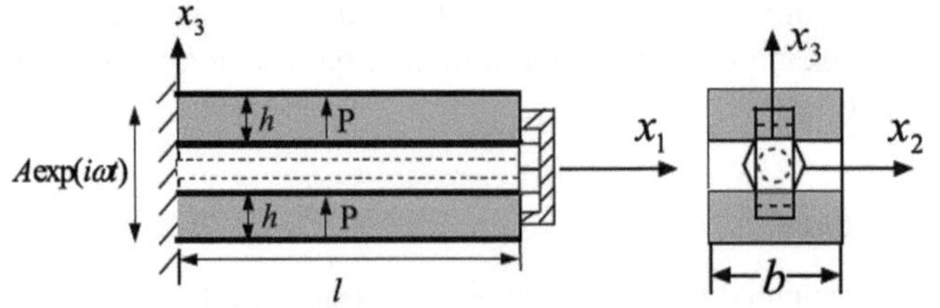

Fig. 1.46 A method to apply axial preload to a piezoelectric bimorph [70]

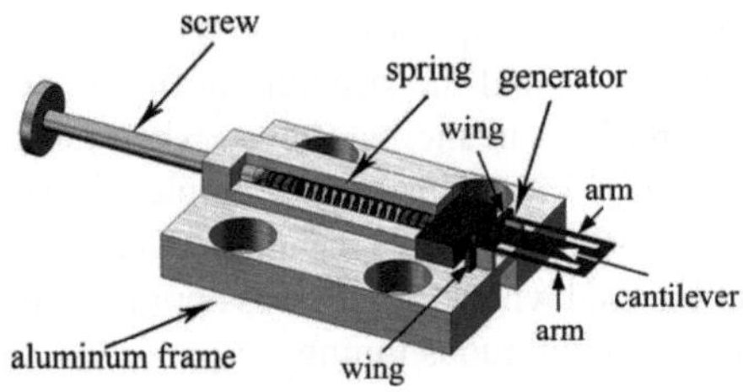

Fig. 1.47 Schematic diagram of the test device [71]

This principle was demonstrated by Eichhorn et al. [71]. Figure 1.47 shows a schematic diagram of the test device. The piezoelectric generator consisted of a piezo-polymer-composite cantilever beam with arms on both sides to enable the application of an axial force to the free end of the beam. The arms were connected to the base with two wings. These wings were used to transmit the force to the arms, which in turn apply the load to the free end of the beam. The tuning force was applied by a screw and a steel spring. The axial load depends linearly on the deflection of the spring, which in turn was proportional to the number of revolutions of the screw. The spring pushes the whole generator base against two blocks of which the counter-pressure generates the pre-stress in the arms and the stabilizing wings. The screw, spring and generator were all mounted on the same aluminium frame. This is another example of manual intermittent tuning.

In tests only a compressive load was applied. Figure 1.48 shows the test results of this generator under vibration level of 63.7 m s^{-2}. It was found that with the increase of compressive load, the resonant frequency, output voltage and the Q-factor reduced. By cutting notches on the wings the tuning efficiency could be increased. With notches in the wings, a resonant frequency shift of more than 20% was achieved with a total force of 22.75 N (Fig. 1.49). The tuning range was from 290 to 380 Hz with compressive load up to 22.75 N. The dimensions of the cantilever are 20 mm × 5 mm × 0.44 mm and the overall width of the device including arms is 13 mm.

A non-contact method of applying axial load to a cantilever-based micro-generator is reported by Zhu et al. [72, 73] who presented a tunable electromagnetic vibration-based micro-generator with closed-loop frequency tuning. Frequency tuning was realized by applying an axial tensile magnetic force to the micro-generator (Fig. 1.50).

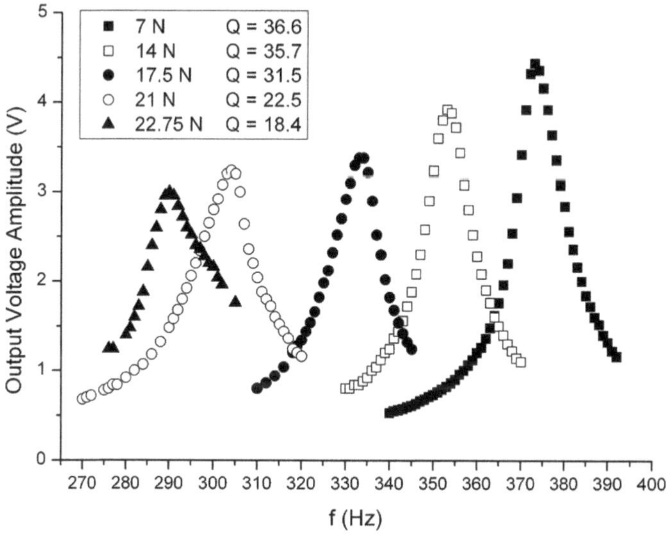

Fig. 1.48 Test results under vibration of 63.7 m s^{-2} [71]

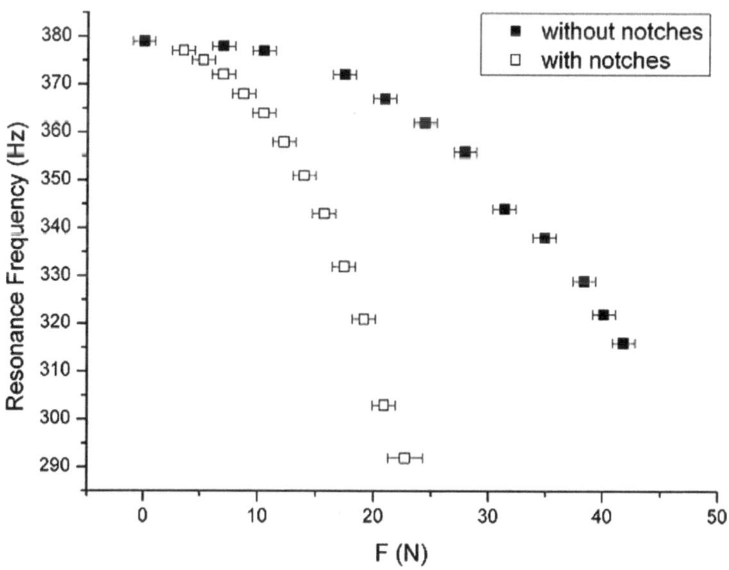

Fig. 1.49 Comparison of tuning efficiency of wings with and without notches [71]

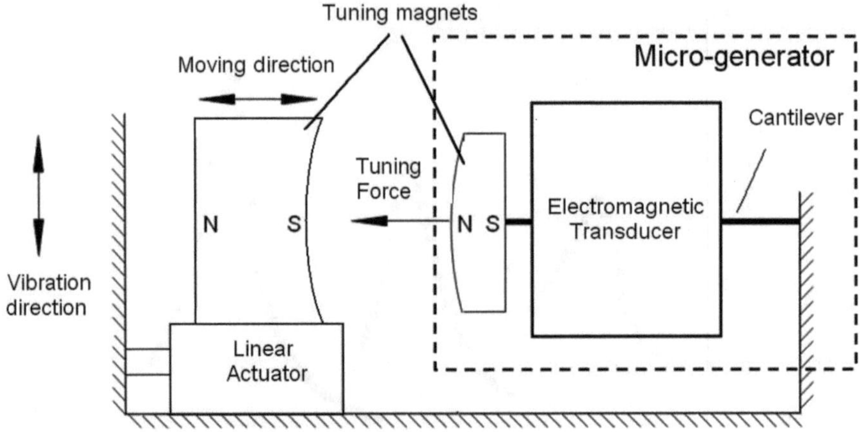

Fig. 1.50 Schematic diagram of the tuning mechanism [72]

The tuning force was provided by the attractive force between two tuning magnets with opposite poles facing each other. One magnet was fixed at the free end of a cantilever while the other was attached to an actuator and placed axially in line with the cantilever. The distance between the two tuning magnets was adjusted by the linear actuator. Thus, the axial load on the cantilever and hence the resonant frequency was changed. The areas where the two magnets face each other were curved to maintain a constant gap between them over the amplitude range of the generator. Figure 1.51 shows the test results of the resonant frequency variation of distance between the two tuning magnets. The tuning range of the proposed micro-generator was from 67.6 to 98 Hz based on the original resonant frequency of 45 Hz by changing the distance between two tuning magnets from 5 to 1.2 mm.

Experimentally, the generator produced a power of 61.6–156.6 μW over the tuning range when it was excited at a constant vibration acceleration level of 0.59 m s^{-2}. It was also found that the tuning mechanism does not affect the damping of the micro-generator over 60% of the tuning range. However, when the tuning force becomes larger than the inertial force caused by vibration, total damping is increased and the output power is less than expected from theory (see resonant peaks at 92 and 98 Hz in Fig. 1.52).

1.5.3 Electrical Tuning Methods

All the frequency tuning methods mentioned above are based on mechanical methods. In this section, methods to tune the resonant frequency of a vibration-based micro-generator electrically will be detailed.

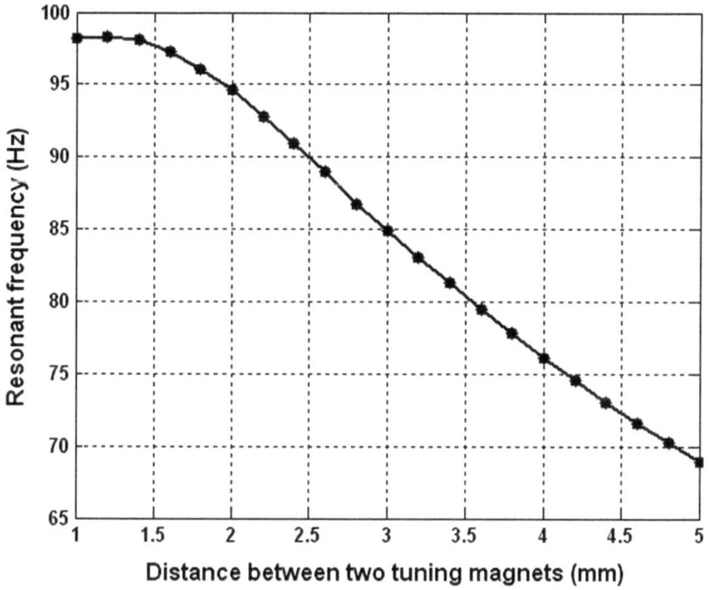

Fig. 1.51 Resonant frequency with variance of distances between two tuning magnets [72]

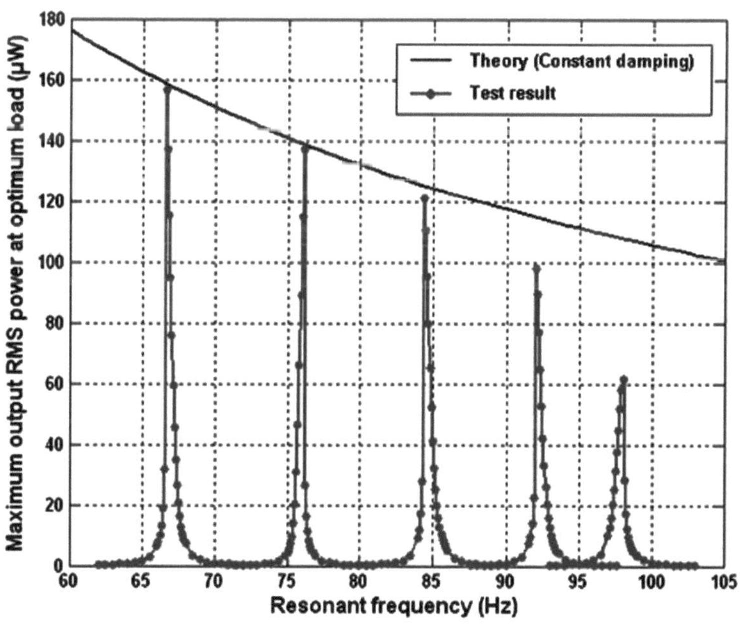

Fig. 1.52 Power spectrum of the micro-generator (excited at 0.59 m s^{-2}) [72]

1.5.3.1 Principles

The basic principle of electrical tuning is to change the electrical damping by adjusting the electrical load. As all reported generators using electrical tuning are piezoelectric, in this section, only the piezoelectric micro-generator will be addressed. As resistive loads reduce the efficiency of power transfer and load inductances are difficult to be varied, it is most feasible to adjust capacitive loads to realize electrical tuning.

Figure 1.53 shows a schematic diagram of a bimorph piezoelectric generator with a mass, m, on the tip. l_b and l_m are the effective length of the cantilever and mass, respectively. w is the width of the cantilever. t_p and t_s are the thickness of the piezoelectric layer and substrate layer, respectively, and t_g is the distance from the centre of the substrate layer to the centre of the piezoelectric layer. Electrodes of the generator have been omitted in Fig. 1.53.

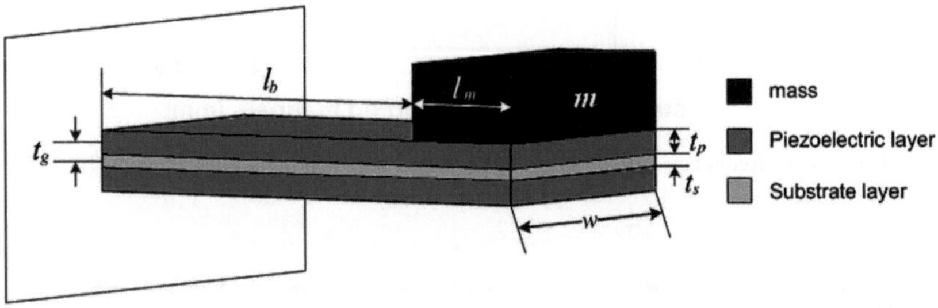

Fig. 1.53 Piezoelectric bimorph generator

This bimorph piezoelectric generator can be represented using an equivalent circuit as shown in Fig. 1.54. L_m, R_m and C_m represent the mass, damping, and spring in the mechanical part, respectively. C_p is the capacitance of the piezoelectric layer, C_L and R_L are the capacitive and resistive loads, respectively. V is the RMS voltage across the resistive load.

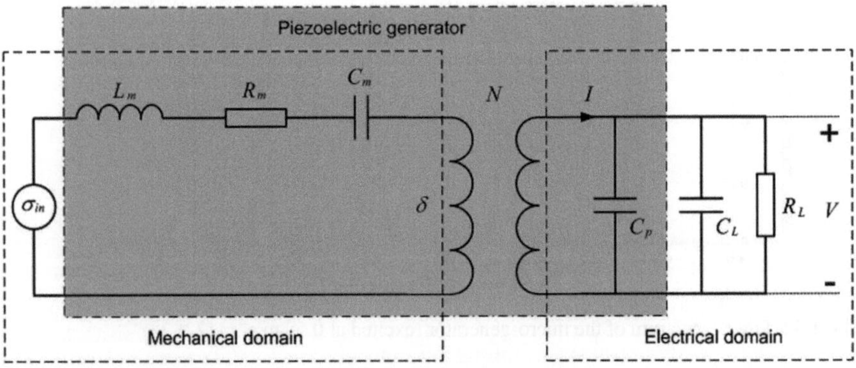

Fig. 1.54 Equivalent circuit of piezoelectric generator with capacitive and resistive loads

The transformer relates the mechanical domain to the electrical domain according to the model of the piezoelectric effect. Specifically, it relates stress (σ) to electric field (E) at zero strain or electrical displacement (D) to strain (δ) at zero electric field. Rewriting equations for piezoelectric effect, which have been described in Eqs. (1.31) and (1.32), leads to the equations for the transformer as

$$D = -d_{31} Y_{\mathrm{p}} \delta \tag{1.62}$$

where d_{31} is the piezoelectric strain coefficient in 31 mode and Y_{p} is Young's modulus of the piezoelectric material.

Hence, the transform ratio N is given by

$$N = -d_{31} Y_{\mathrm{p}} \tag{1.63}$$

Equation (1.64) can be derived to present the mechanical dynamics of the system with electrical coupling. Detailed derivation of this model can be found in [21].

$$\Delta(s^2 + 2\zeta \omega_{\mathrm{r}} s + \omega^2) = \frac{\omega_{\mathrm{r}}^2 d_{31} a}{2t_{\mathrm{c}}} V + b^* A_{\mathrm{in}} \tag{1.64}$$

where Δ is Laplace transform of strain, δ, A_{in} is the vibration acceleration, ζ is the damping factor, ω_{r} is the untuned resonant frequency, $a = 1$ if the two piezoelectric layers are connected in series and $a = 2$ if they are connected in parallel and s is the Laplace variable. b^* is given by

$$b^* = \frac{3t_{\mathrm{g}}}{l_{\mathrm{b}}^2} \cdot \frac{2l_{\mathrm{b}} + l_{\mathrm{m}} - l_{\mathrm{e}}}{2l_{\mathrm{b}} + \frac{3}{2} l_{\mathrm{m}}} \tag{1.65}$$

where l_{e} is the length of the electrodes.

Furthermore, analysis in the electrical domain gives the following equation:

$$\Delta = \left(s + \frac{1}{R_{\mathrm{L}} C_{\mathrm{pL}}}\right) \frac{V C_{\mathrm{pL}}}{s \Sigma} \tag{1.66}$$

where $C_{\mathrm{pL}} = C_{\mathrm{p}} + C_{\mathrm{L}}$ and $\Sigma = -a \cdot d_{31} \cdot Y_{\mathrm{c}} \cdot l_{\mathrm{e}} \cdot w$. l_{e} is the length of the electrodes.

Combining Eqs. (1.64) and (1.66) gives the transfer function of the system as

$$\frac{V}{A_{\mathrm{in}}} = \frac{s \Sigma b^*}{s^3 + \left(\frac{1}{R_{\mathrm{L}} C_{\mathrm{pL}} + 2\zeta \omega_{\mathrm{r}}}\right) s^2 + \left(\omega_{\mathrm{r}}^2 + \frac{2\zeta \omega_{\mathrm{r}}}{R_{\mathrm{L}} C_{\mathrm{pL}}} - \frac{a \Sigma \omega_{\mathrm{r}}^2 d_{31}}{2t_{\mathrm{c}}}\right) s + \frac{\omega_{\mathrm{r}}^2}{R_{\mathrm{L}} C_{\mathrm{pL}}}} \tag{1.67}$$

which leads to the expression of the voltage across the resistive load given by

$$V(\omega) = \frac{j\omega\Sigma b^* A_{\text{in}}}{\left[\frac{\omega_\text{r}^2}{R_\text{L}} - \left(\frac{1}{R_\text{L}} + 2\zeta\omega_\text{r} C_{\text{pL}}\right)\omega^2\right] + j\omega\left[(\omega_\text{r}^2 - \omega^2)C_{\text{pL}} + \frac{2\zeta\omega_\text{r}}{R_\text{L}} - \frac{a\Sigma\omega_\text{r}^2 d_{31}}{2t_\text{c}}\right]} \tag{1.68}$$

The power in the resistive load is given by

$$P(\omega) = \frac{V(\omega)^2}{R_\text{L}}$$
$$= \frac{1}{R_\text{L}} \cdot \frac{(\omega\Sigma b^* A_{\text{in}})^2}{\left[\frac{\omega_\text{r}^2}{R_\text{L}} - \left(\frac{1}{R_\text{L}} + 2\zeta\omega_\text{r} C_{\text{pL}}\right)\omega^2\right]^2 + \omega^2\left[(\omega_\text{r}^2 - \omega^2)C_{\text{pL}} + \frac{2\zeta\omega_\text{r}}{R_\text{L}} - \frac{a\Sigma\omega_\text{r}^2 d_{31}}{2t_\text{c}}\right]^2} \tag{1.69}$$

It is known that $f(x, y) = x^2 + y^2 \le 2xy$ and that $f(x, y)$ becomes a minimum only if $x = y$ (i.e. $\frac{1}{f(x,y)}$ is maximum only if $x = y$). Therefore, Eq. (1.69) reaches maximum when

$$\frac{\omega_\text{r}^2}{R_\text{L}} - \left(\frac{1}{R_\text{L}} + 2\zeta\omega_\text{r} C_{\text{pL}}\right)\omega^2 = \omega\left[(\omega_\text{r}^2 - \omega^2)C_{\text{pL}} + \frac{2\zeta\omega_\text{r}}{R_\text{L}} - \frac{a\Sigma\omega_\text{r}^2 d_{31}}{2t_\text{c}}\right] \tag{1.70}$$

Rearranging Eq. (1.70) leads to a cubic function of the form

$$\omega^3 + X\omega^2 + Y\omega + Z = 0 \tag{1.71}$$

where

$$X = -\left(\frac{1}{R_\text{L} C_{\text{pL}}} + 2\zeta\omega_\text{r}\right)$$
$$Y = -\left(\omega_\text{r}^2 + \frac{2\zeta\omega_\text{r}}{R_\text{L} C_{\text{pL}}} - \frac{a\Sigma\omega_\text{r}^2 d_{31}}{2t_\text{c} C_{\text{pL}}}\right)$$
$$Z = \frac{\omega_\text{r}^2}{R_\text{L} C_{\text{pL}}}$$

The real solution of Eq. (1.71) gives the function of resonant frequency with respect to the load capacitance as

$$\omega(C_\text{L}) = \frac{\sqrt[3]{\Omega + 12\sqrt{\Psi}}}{6} - \frac{2(Y - \frac{X^2}{3})}{\sqrt[3]{\Omega + 12\sqrt{\Psi}}} - \frac{X}{3} \tag{1.72}$$

where $\Omega = 36XY - 108Z - 8X^3$ and $\Psi = 12Y^3 - 3X^2Y^2 - 54XYZ + 81Z^2 + 12X^3Z$.

Equations (1.69) and (1.72) indicate that output power and the resonant frequency of a bimorph piezoelectric generator vary with variations of the load capacitance. Figure 1.55 compares the resonant frequencies and power output of electrically tunable piezoelectric generators of different piezoelectric materials with

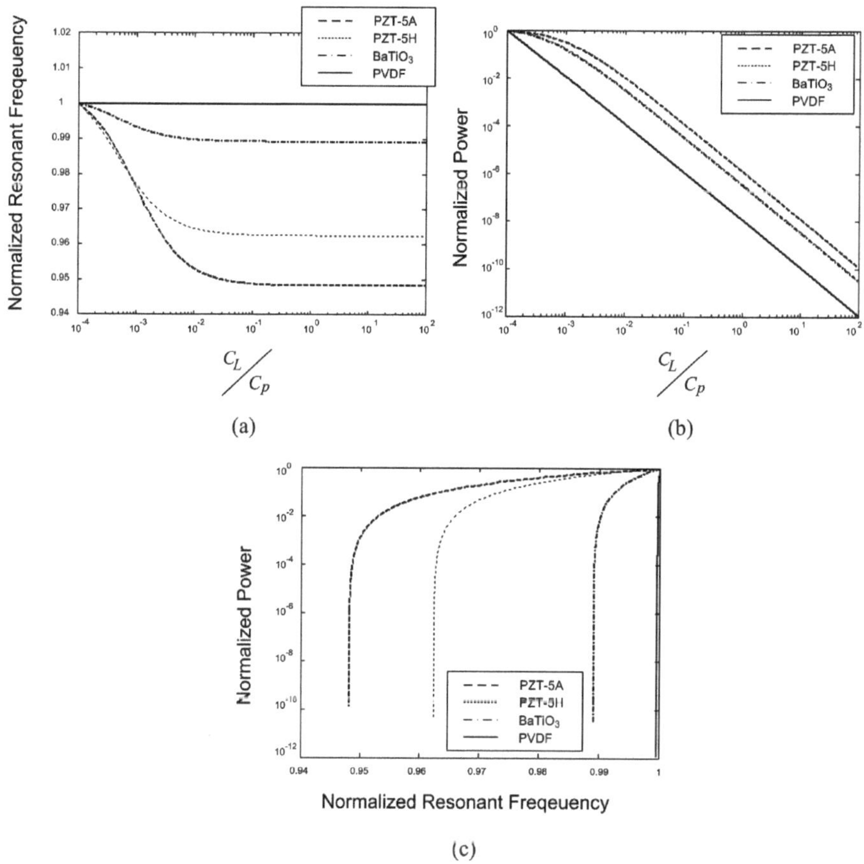

Fig. 1.55 Performance of a piezoelectric generator with different piezoelectric materials: (**a**) Resonant frequency (**b**) output power; and (**c**) output power versus resonant frequency

variation of load capacitances. These generators are identical except for the piezoelectric material. The coefficients used in the simulation are listed in Table 1.2.

The resonant frequency as well as the output power reduces with increasing load capacitance. It was found that PZT-5H is the best of these four piezoelectric materials for an electrically tunable piezoelectric generator. Important considerations relating to the tunability of the piezoelectric generator are as follows:

- The material of the substrate layer and mass does not affect the tunability.
- A piezoelectric material with higher Young's modulus, strain coefficient and smaller permittivity provides a larger tuning range.
- The ratio of the thickness of the piezoelectric layer to the thickness of the substrate layer should be small to increase the tuning range.
- The capacitance of the piezoelectric layer should be minimized to increase the tuning range.

- If both piezoelectric layers are used for tuning, connection of these two layers in parallel gives a larger tuning range than connection in series.
- The total damping should be kept low to increase the tuning range.

1.5.3.2 Examples of Electrically Tunable Micro-generators

Wu et al. [74] used this method to tune the resonant frequency of a generator composed of a piezoelectric bimorph cantilever. The upper piezoelectric layer was used for frequency tuning while the lower layer was used for energy harvesting. The tunable bandwidth of this generator was 3 Hz between 91.5 and 94.5 Hz. The charging time of the generator was compared with and without the tuning system. Experimentally, it was found that, when the device was excited under random frequencies from 80 to 115 Hz, the average harvesting output power of the generator with tuning was about 27.4% higher than that without tuning and the charging time was shortened by using tuning system. These results showed a significant improvement of average harvested power output by using an electrical tuning method.

Charnegie [75] presented another piezoelectric micro-generator based on a bimorph structure and adjusted its load capacitance. Again, one piezoelectric layer was designed for energy harvesting while the other is used for frequency tuning (Fig. 1.56).

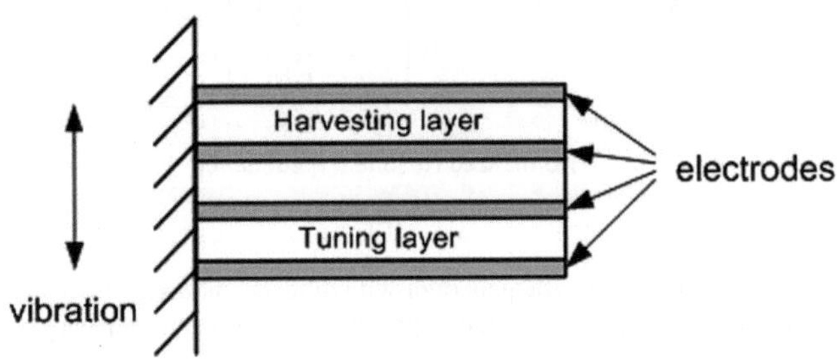

Fig. 1.56 Piezoelectric bimorph used for electrical frequency tuning

The test results showed that if only one layer was used for frequency tuning (Fig. 1.57a), the resonant frequency can be tuned an average of 4 Hz with respect to the untuned frequency of 350 Hz, i.e. 1.14% tuning by adjusting the load capacitance from 0 to 10 mF (Fig. 1.58a). If both layers were used for frequency tuning (Fig. 1.57b), the tuning range was an average of 6.5 Hz, i.e. 1.86% of tuning by adjusting the same amount of the load capacitance (Fig. 1.59a). It was found that if one layer was used for tuning and the other for energy harvesting (Fig. 1.57a), the output power did not reduce with the increase of the load capacitance (Fig. 1.58b). However, if both frequency tuning and energy harvesting were achieved using the same layer (Fig. 1.57b), the output power decreased when the load capacitance became larger (Fig. 1.59b).

Cammarano et al. [76] presented a vibration-based energy harvester whose resonant frequency can be tuned by varying its electrical load (containing both real

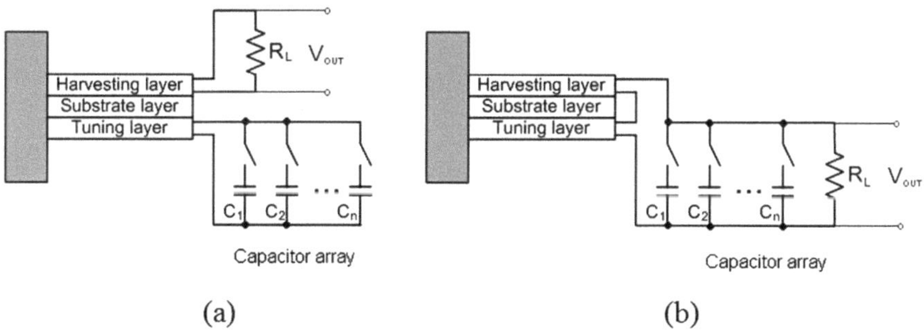

Fig. 1.57 Frequency tuning and energy harvesting using (**a**) the same layers and (**b**) different layers

and reactive impedances). It was found experimentally that the −3 dB (half-power) bandwidth of the energy harvester (16 Hz) is over three times greater when presented with an optimized load impedance compared to that for the same harvester presented with an optimized resistive-only load (4.5 Hz). They also developed an analytical model of the system. Readers may refer to their paper for more details if interested.

1.6 Strategies to Widen Bandwidth

The other commonly used solution to increase the operational frequency range of kinetic energy harvesters is to widen the bandwidth. To date, strategies to widen the bandwidth include using a generator array consisting of small generators with different resonant frequencies, introducing an amplitude limiter to the device, using coupled oscillators, employing non-linear and bi-stable structures and designing a large generator with a large inertial mass and high degree of damping. In this section, details of generator array, amplitude limiter and non-linear and bi-stable structures will be covered. The strategy of employing a single large generator will not be detailed as it can be simply described using Eq. (1.14) while it will be considered in the comparison of different strategies later in Section 1.7.

1.6.1 Generator Array

A generator array consists of multiple small generators, each of which has different dimensions and mass and hence different resonant frequencies (Fig. 1.60). Thus, the assembled generator has a wide operational frequency range while the Q-factor does not decrease. Figure 1.61 shows the power spectrum of a generator array which is a combination of the power spectra of each small generator. The operational frequency band of the generator is thus essentially increased. The drawback of this approach is the added complexity of fabricating an array of generators and the increased total volume of the device depending upon the number of devices in the array.

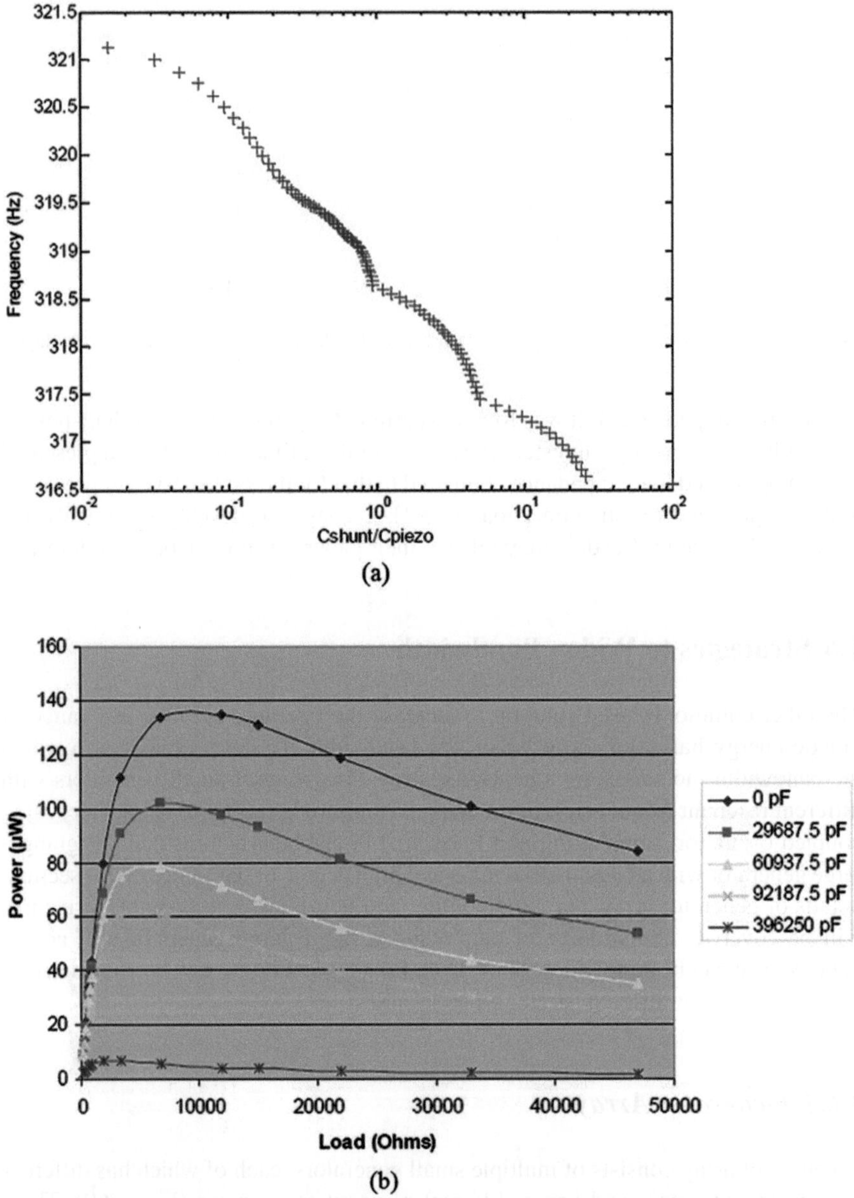

Fig. 1.58 (**a**) Resonant frequency and (**b**) output power versus load capacitance while tuning and energy harvesting in different layers [75]

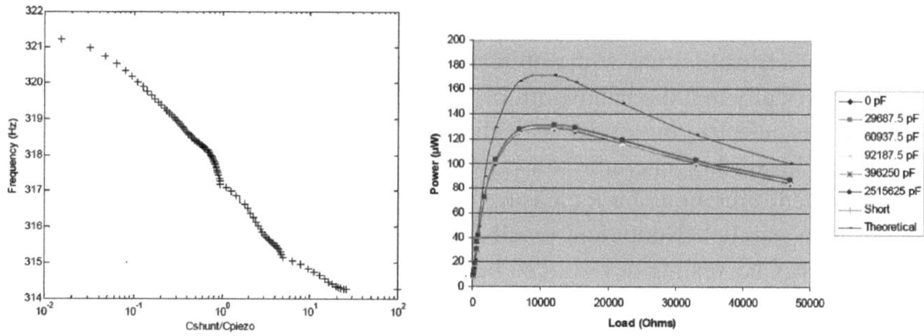

(a) (b)

Fig. 1.59 (**a**) Resonant frequency and (**b**) output power versus load capacitance while tuning and energy harvesting in same layers [75]

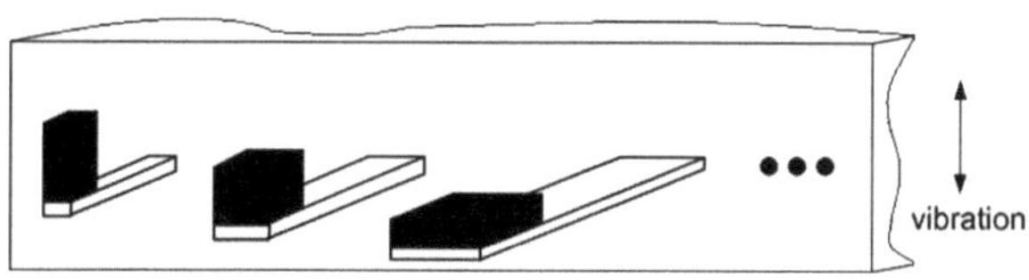

Fig. 1.60 A mechanical band-pass filter with a set of cantilever beams

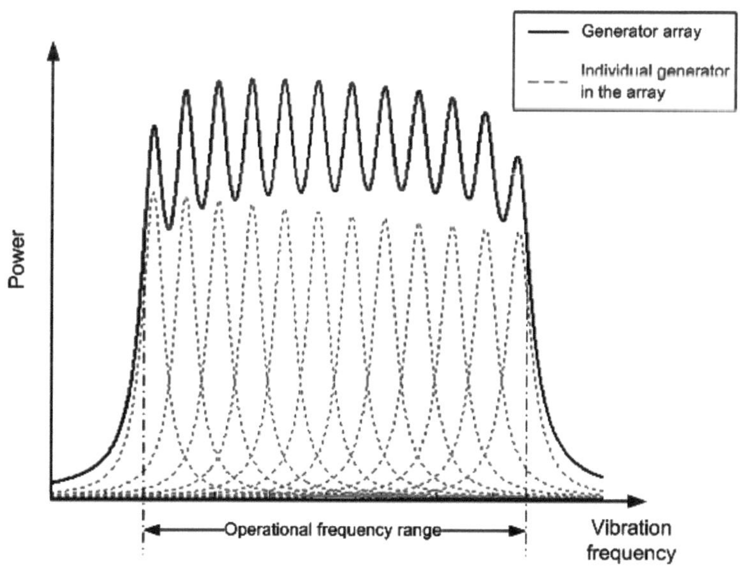

Fig. 1.61 Power spectrum of a generator array

Xue et al. [77] presented a broadband piezoelectric harvesters consisting of multiple piezoelectric bimorphs with different thicknesses of piezoelectric layers. It was found analytically that the bandwidth of a generator can be widened by connecting multiple piezoelectric bimorphs with different dimensions in series. In addition, the bandwidth of the generator can be shifted to the dominant frequency domain of the ambient vibrations by increasing or decreasing the number of piezoelectric bimorphs in parallel. Numerical results showed that the bandwidth of the piezoelectric energy harvesting devices can be tailored by the connection patterns (i.e. in series and in parallel) among piezoelectric bimorphs (Figs. 1.62 and 1.63).

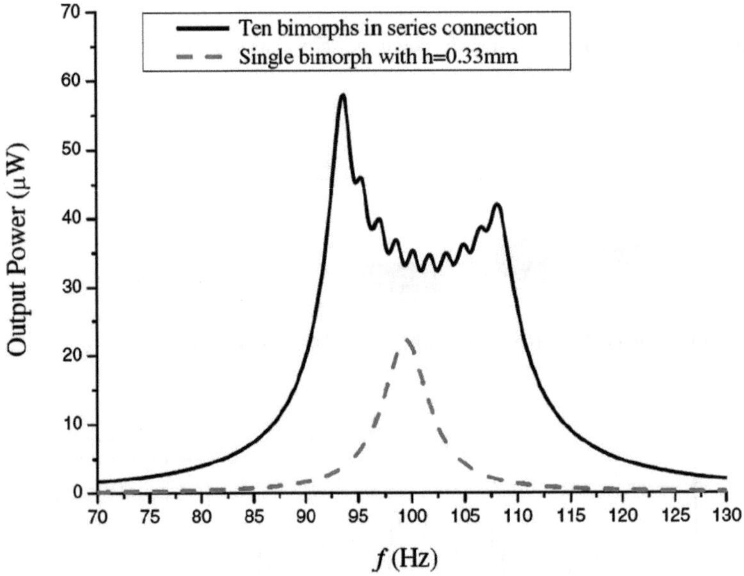

Fig. 1.62 Comparison of power spectrum for a single piezoelectric bimorph and 10 piezoelectric bimorphs in series with various thicknesses of piezoelectric layer [77]

Feng and Hung [78] presented a micromachined piezoelectric generator with a wide bandwidth. The device was designed to achieve an optimal figure of merit (FOM) which is defined as (bandwidth)$^2\times$ (the maximum displacement of cantilever structures under a given acceleration under static conditions). The generator consisted of four cantilever structures connected in parallel and has dimensions of 3 mm $\times$ 3 mm $\times$ 5 mm. These cantilevers had different masses or centre of gravity and hence different resonant frequencies (Fig. 1.64). The designed generator was targeted at producing microwatts to milliwatts in a wide mechanical vibration range from 300 to 800 Hz (Fig. 1.65) but no test results were reported to date.

A multifrequency piezoelectric generator intended for powering autonomous sensors from background vibrations was presented by Ferrari et al. [79]. The generator consisted of multiple bimorph cantilevers with different natural frequencies of which the rectified outputs were fed to a single storage capacitor. A generator

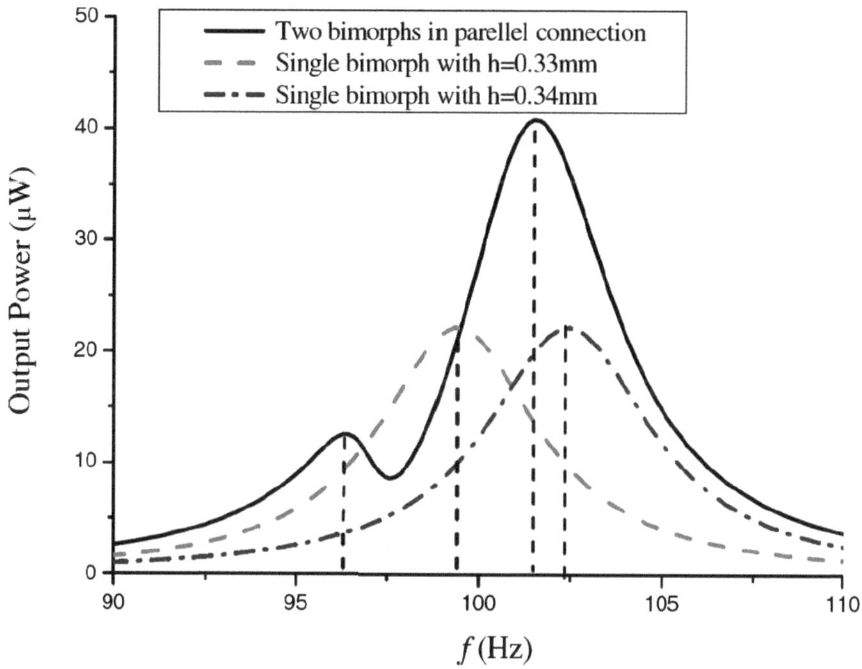

Fig. 1.63 Effect of piezoelectric bimorphs in parallel on harvester performance [77]

with three commercially available piezoelectric bimorph cantilevers was examined. Each cantilever has the same dimensions of 15 mm × 1.5 mm × 0.6 mm and different masses, 1.4, 0.7, and 0.6 g, respectively. The generator was used to power a battery-less sensor module. It was concluded that a generator array operating with wideband frequency vibrations provides improved overall energy conversion over a single generator at the expense of larger volume.

Sari et al. [80] reported a micromachined electromagnetic generator with a wide bandwidth. The generator consists of a series of cantilevers with various lengths and hence resonant frequencies (Fig. 1.66). These cantilevers are distributed in a 12.5 mm × 14 mm area. The length of the cantilevers increased gradually so that the cantilevers have overlapping frequency spectra with the peak powers at similar but different frequencies. This resulted in a widened bandwidth as well as an increase in the overall output power. Experimentally, the device generated 0.5 μW continuous power at 20 mV voltage between 3.3 and 3.6 kHz of ambient vibration. Figure 1.67 shows the power spectrum of this generator.

Lin et al. [81] reported a multi-cantilever piezoelectric MEMS generator, which has the ability to scavenge mechanical energy of ambient vibrations and transforms it into useful electrical power. The generator comprises four cantilever-type devices, two mode 31 devices and two mode 33 devices, which were made by a silicon process in a single die. The four cantilever devices can be connected in series or in parallel so as to possess different output characteristics. The measurement results

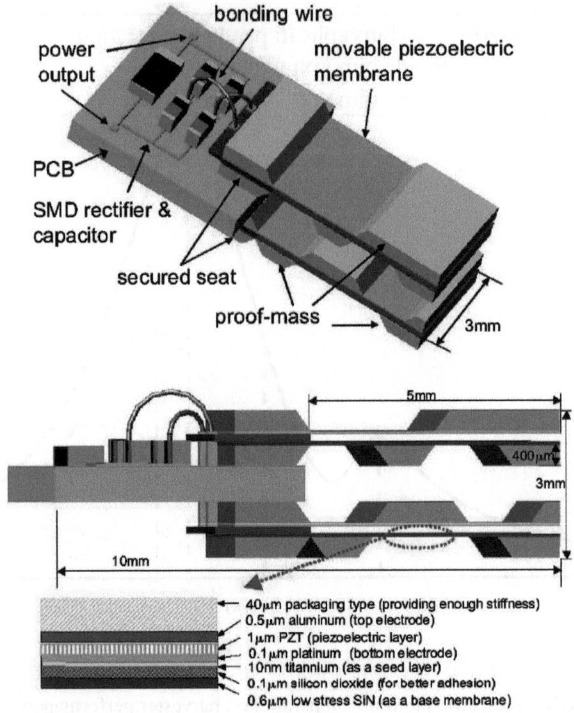

Fig. 1.64 Conceptual diagram of the piezoelectric wide bandwidth micro-generator [78]

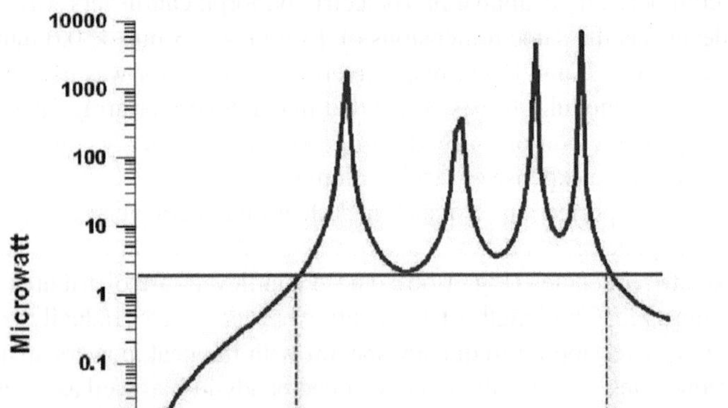

Fig. 1.65 Estimated power generation with the power range of microwatts to milliwatts in a wide bandwidth [78]

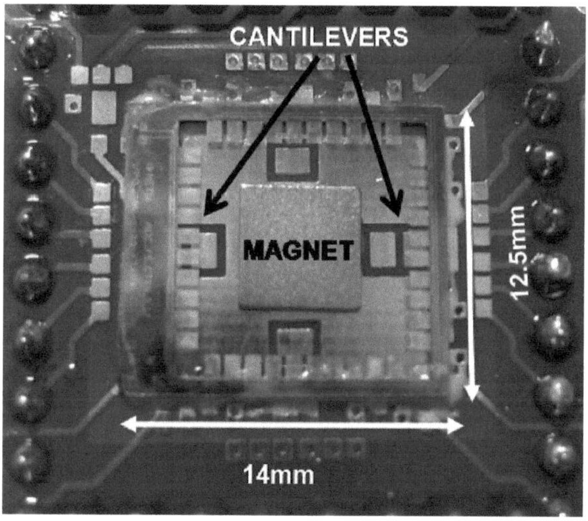

Fig. 1.66 Photograph of a wideband electromagnetic generator [80]

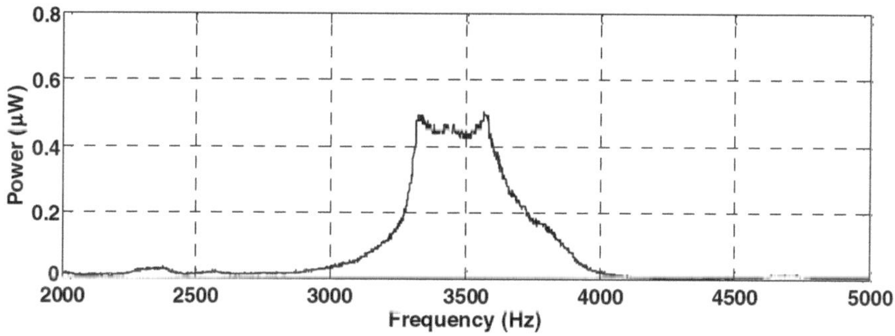

Fig. 1.67 Power spectrum of Sari's generator [80]

show that the prototype device possesses resonance frequencies between 237 and 244.5 Hz. However, no information of output power has been given.

1.6.2 Amplitude Limiter

Another method of increasing the bandwidth of a vibration-based micro-generator was reported by Soliman et al. [82, 83]. The bandwidth of the device was increased by using a mechanical stopper (amplitude limiter) to limit the amplitude of the resonator (Figs. 1.68 and 1.69). The theory behind this method is complex and details can be found in [82]. This method can increase the bandwidth of the generator when the excitation frequency was gradually increased. However, the bandwidth remained the same when excitation frequency was gradually reduced. Experimental measurements showed that the bandwidth was 240% wider than that of the architecture without a stopper at the half-power level but the maximum output voltage was 30% less (Fig. 1.70) in the case when the excitation frequency was increased. The dimensions of the cantilever are 45.3 mm × 10 mm × 1.02 mm and the mass is extrapolated to be 2.92 g. It should be noted that since this principle relies on continuous physical contact with the cantilever, it is unlikely to provide a reliable long-term solution for increasing bandwidth.

1.6.3 Coupled Oscillators

The method of widening the operational bandwidth of the MEMS generator using coupled oscillators was reported by Petropoulos et al. [84]. The proposed generator has a pair of coupled oscillators that consist of two springs, two masses and two dampers. The first spring connects the inertial frame and the first mass while the second spring connects the two mass while each mass has a damper to the frame as shown in Fig. 1.71. The analytic model shows that this type of generators has flat response for power generation over a wider frequency range. However, the maximum output power of the generator is significantly decreased than that of the generator with one mass (Fig. 1.72).

1.6.4 Non-linear Generators

The theory of vibration energy harvesting using non-linear generators was investigated by Ramlan et al. [85]. Instead of using conventional second-order model as Eq. (1.1), non-linear generators were modelled using Duffing's equation as follows:

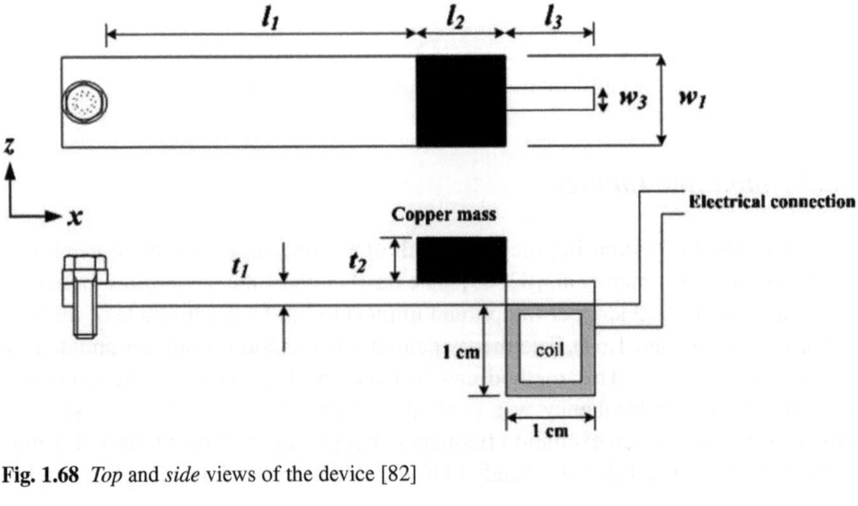

Fig. 1.68 *Top* and *side* views of the device [82]

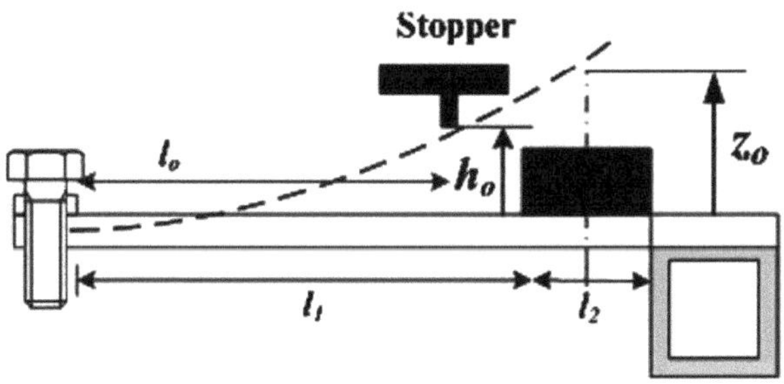

Fig. 1.69 Increase the bandwidth using an amplitude limiter [82]

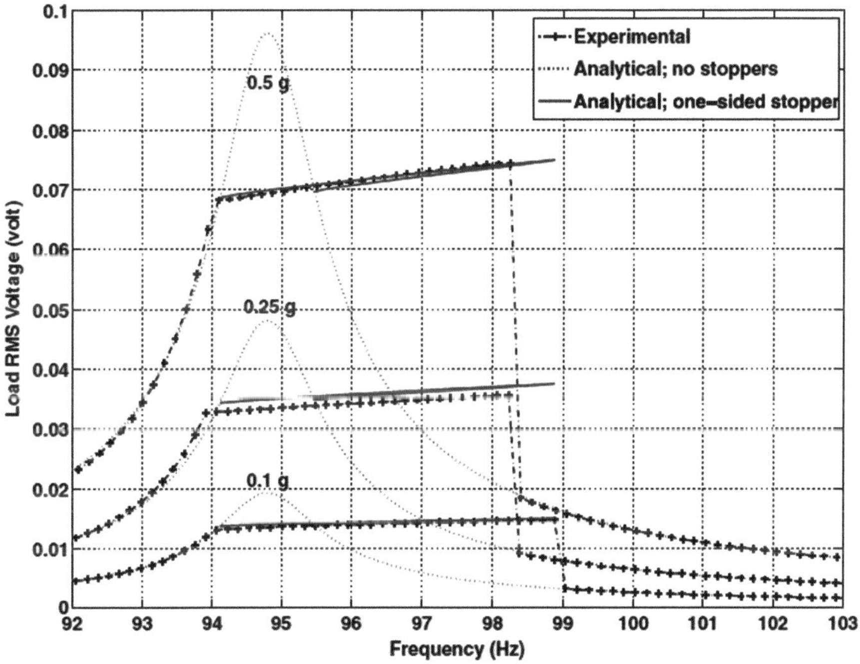

Fig. 1.70 Voltage on load versus excitation frequency [82]

$$m \cdot \frac{d^2 z(t)}{dt^2} + b \cdot \frac{dz(t)}{dt} + k \cdot z(t) + k_n [z(t)]^3 = -m \cdot \frac{d^2 y(t)}{dt^2} \tag{1.73}$$

where the spring force is the combination of linear force,

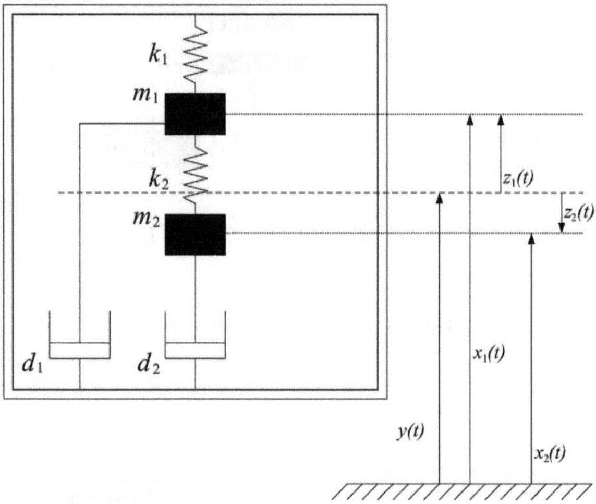

Fig. 1.71 Schematic model of a coupled oscillator [84]

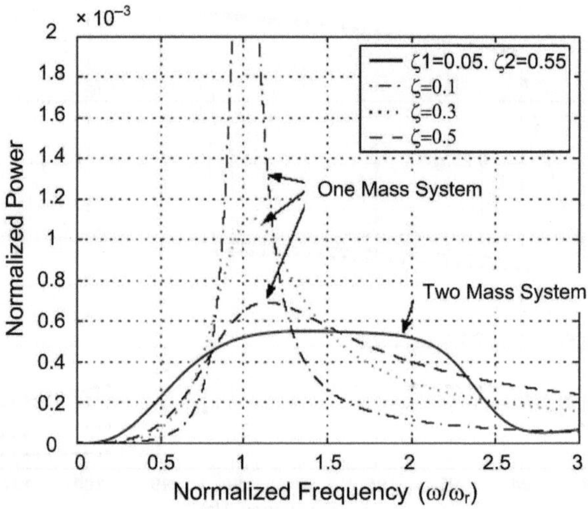

Fig. 1.72 Power spectrum for optimized two-mass system and for equivalent one-mass system with various ζ values [84]

$$\Omega_a = \sqrt{\frac{3}{4}\alpha Z^2 + (1 - 2\zeta^2) - \frac{\sqrt{a^2 - 3\alpha\zeta^2 Z^4 + 4\zeta^2 Z^2(\zeta^2 - 1)}}{Z}} \tag{1.74}$$

$$\Omega_b = \sqrt{\frac{3}{4}\alpha Z^2 + (1 - 2\zeta^2) + \frac{\sqrt{a^2 - 3\alpha\zeta^2 Z^4 + 4\zeta^2 Z^2(\zeta^2 - 1)}}{Z}} \tag{1.75}$$

where α is proportional to the non-linear spring factor, k_n, Z is the amplitude of the proof mass, ζ is the damping factor and a is the normalized excitation acceleration.

Such devices have a hardening spring which has the effect of shifting the resonant frequency. Numerical and analytical studies showed that a device with a hardening spring has a larger bandwidth over which power can be harvested due to the shift in the resonance frequency. Analytical results also showed that the bandwidth of the hardening system depends on the damping ratio, the non-linearity and the input acceleration (Fig. 1.73). Ideally, the maximum amount of power harvested by a non-linear system with a hardening stiffness is the same as the maximum power harvested by a linear system. The maximum power occurs at a different frequency depending on the non-linearity. It is important to mention that the output power and bandwidth of the non-linear generators depend on the direction of approach of the vibration frequency to the resonant frequency. For a hard non-linearity, bandwidth only increases when approaching the device-resonant frequency from a lower frequency while for a soft non-linearity, bandwidth only increases when approaching the device-resonant frequency from a higher frequency. It is unlikely that these conditions can be guaranteed in real applications.

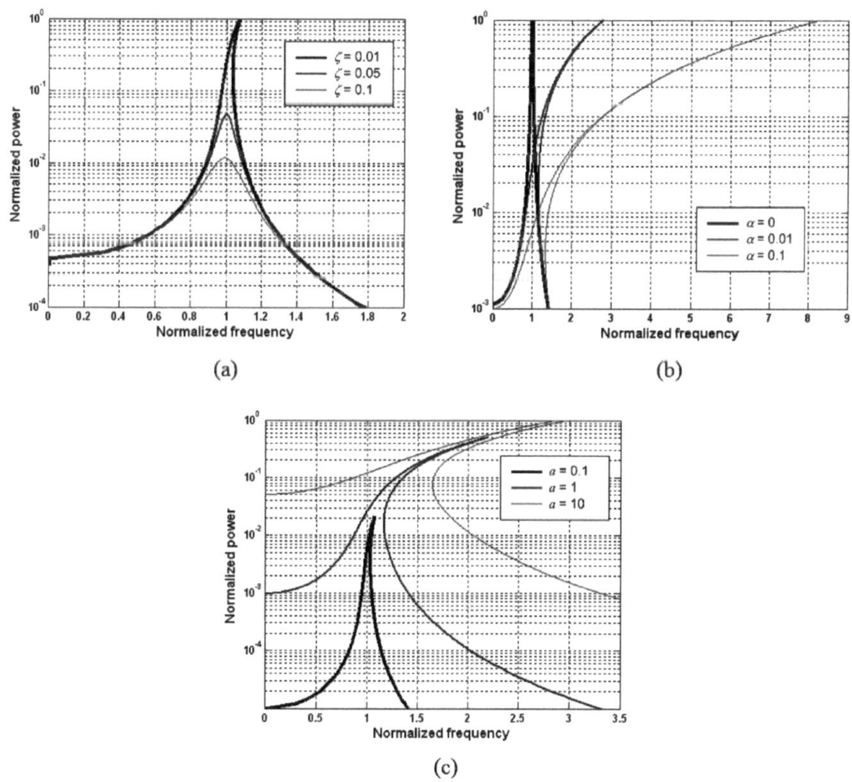

Fig. 1.73 Power spectrum of non-linear generators: (**a**) various damping ratio; (**b**) various non-linearity; and (**c**) various input acceleration

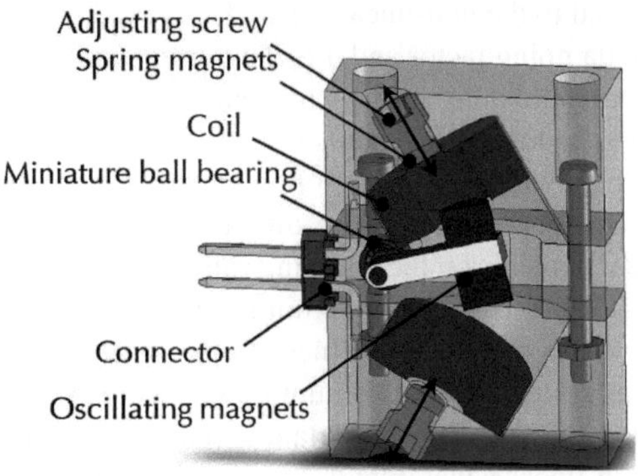

Fig. 1.74 Half-section of the device [86]

Non-linear generators can be realized by replacing using a conventional spring with a magnetic spring. Spreemann et al. [86] reported a tunable electromagnetic vibration energy harvester with a magnetic spring, which combined a tuning mechanism with the non-linear structure. Instead of using a linear suspension, this device was implemented using a rotary suspension (Fig. 1.74). The use of magnetic spring magnets resulted in a non-linear restoring force. As shown experimentally in Fig. 1.75, the resonant frequency shifted by about 30 Hz for a displacement of 1.5 mm of each spring magnet. The maximum output decreased with the increase of the magnet spacing, i.e. as the resonant frequency decreased. Also the bandwidth of the device increased as the space between magnets became smaller, i.e. non-linearity increased. This agrees with the analysis result shown in Fig. 1.73. The generator has a volume of approximately 2.5 cm^3.

In addition, the design and analysis of an energy harvesting device with magnetic restoring forces to levitate an oscillating centre magnet was presented by Mann and Sims [87]. Figure 1.76 shows the schematic diagram of the device. The device has two magnets that were mechanically attached to the generator housing. A centre magnet was placed between the two fixed magnets and the magnetic poles were oriented to repel the centre magnet, thus suspending the centre magnet with a non-linear restoring force. The non-linearity allows the linear resonance to be tuned by simply changing the spacing between outer and centre magnets.

It was found theoretically and experimentally that the response for both linear and non-linear systems scales almost linearly within some regimes of excitation amplitudes (Fig. 1.77a). However, once the non-linearities have been sufficiently engaged, as shown in Fig. 1.77b, the peak response of the non-linear system does not occur at its linear resonant frequency. In the frequency response for the non-linear system, relatively large amplitudes persist over a much larger range of frequencies, which could prove beneficial for applications with either fixed or varying excitation inputs. Furthermore, the maximum output power of such devices is delivered to the electrical load at a frequency away from linear resonance.

Burrow et al. [88, 89] reported a non-linear generator that consisted of a linear spring with the non-linearity caused by the addition of magnetic reluctance

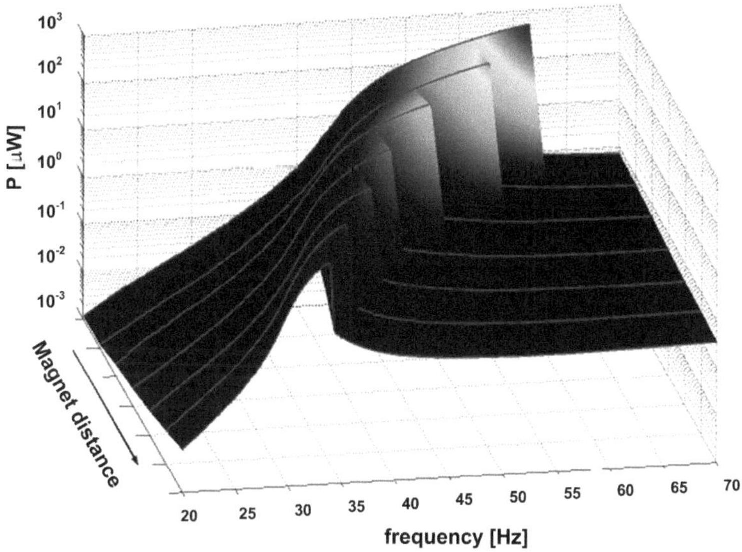

Fig. 1.75 Measured output power [86]

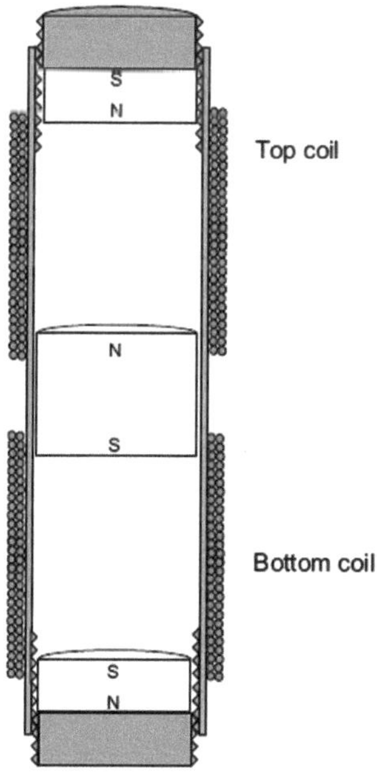

Fig. 1.76 Schematic diagram of magnetically levitated generator [87]

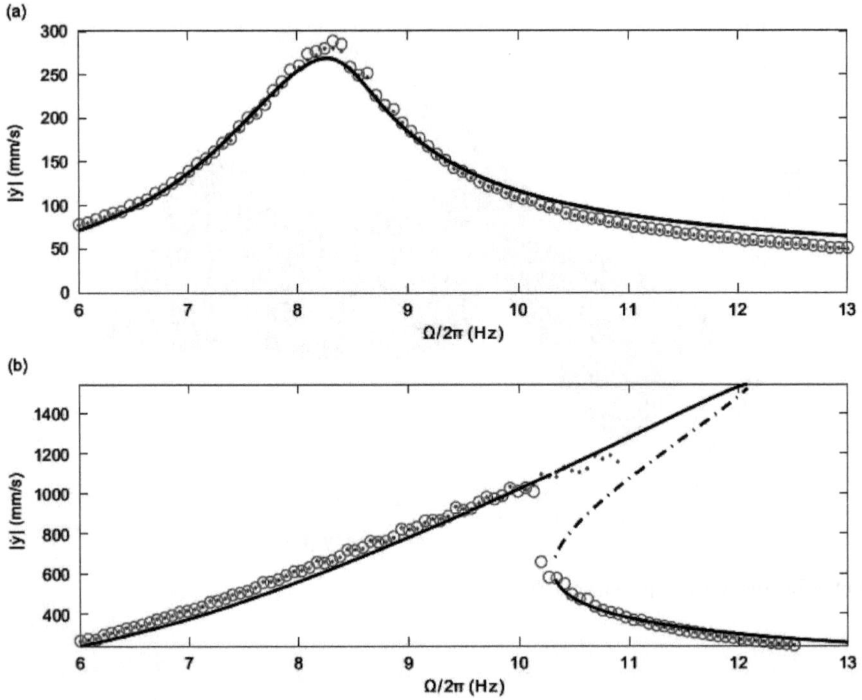

Fig. 1.77 Experimental velocity response amplitudes from forward (*dots*) and reverse frequency sweeps (*circles*) are compared with theory. Theoretical predictions are separated into stable solutions (*solid line*) and unstable solutions (*dashed line*): (**a**) Excitation level of m s^{-2} (**b**) excitation level of 8.4 and m s^{-2} [87]

forces. Figure 1.78 shows the schematic diagram of the non-linear generator. The flux concentrator guides the magnetic flux through the coil. The vibration of the magnets causes a change in direction of the magnetic flux, which induces a voltage across the coil. The reluctance force between the magnets and the flux concentrator resulted in the non-linearity. It is found experimentally that the generator has a wider bandwidth during an up-sweep, i.e. when the excitation frequency was gradually

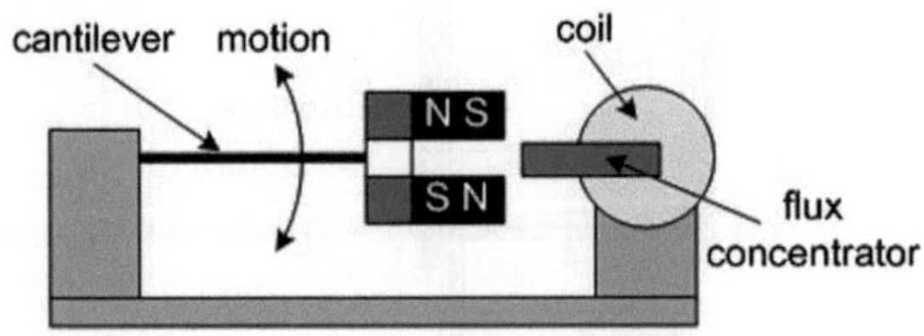

Fig. 1.78 Schematic diagram of a non-linear generator (after [89])

increased while the bandwidth was much narrower during a down-sweep, i.e. when the excitation frequency was gradually decreased.

Tvedt et al. [90] studied non-linear behaviour in an electrostatic vibration energy harvester. The measured non-linear phenomena were described by a lumped model with a non-linear beam displaying both spring softening and hardening. Experimental results show that considerable bandwidth enhancements can be achieved by using non-linear springs without relying on mechanical stopper impacts, resonance tuning or large electromechanical coupling.

1.6.5 Bi-stable Generators

Ramlan et al. [85] studied a bi-stable structure for energy harvesting (also termed the snap-through mechanism). These structures employ a negative stiffness which has the effect of steepening the displacement response of the resonator as a function of time resulting in a higher velocity for a given input excitation. Analysis revealed that the amount of power harvested by a bi-stable device is $4/\pi$ greater than that of the tuned linear device provided the device produces a square wave output for a given sinusoidal input. Numerical results also showed that more power is harvested by the mechanism if the excitation frequency is much less than the generator's resonant frequency. Although the bi-stable mechanism cannot produce a square wave like response under all operating conditions, it offers better performance than the linear mechanism at lower frequencies than the resonant frequency of the linear device. Bi-stable devices also have the potential to cope with mismatch between resonant frequency and vibration frequency.

Galchev et al. [91] reported an electromagnetic generator with a bi-stable structure for scavenging low-frequency non-periodic vibrations. The bi-stable mechanical structure is used to initiate high-frequency mechanical oscillations in an electromagnetic transducer. The fabricated device generated a peak power of 288 μW and an average power of 5.8 μW from an input acceleration of 9.8 m s^{-2} at 10 Hz. The device operates over a frequency range of 20 Hz. The internal volume of the generator is 2.1 cm^3 (3.7 cm^3 including casing), half of a standard AA battery.

Mann and Owens [92] investigated a non-linear energy harvester that uses magnetic interactions to create an inertial generator with a bi-stable potential well. Both theoretical and experimental results show that the potential well escape phenomenon can be used to broaden the frequency response of an energy harvester.

Both Ferrari et al. [93] and Stanton et al. [94] studied the bi-stable structure as shown in Fig. 1.79. Both generators have piezoelectric cantilevers and were tested in vibration with random frequencies. Their experimental results show that bi-stable structure can increase output power of the generator in a wider frequency range.

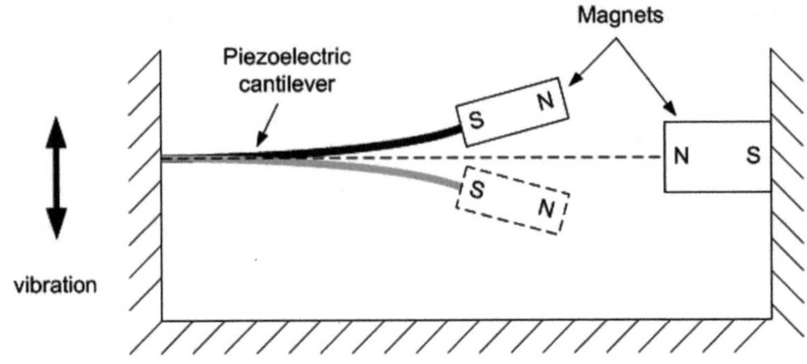

Fig. 1.79 A bi-stable piezoelectric cantilever

1.7 Comparisons of Different Strategies for Adaptive Kinetic Energy Harvesting

It has been proven theoretically and experimentally that both tuning resonant frequency and widening the bandwidth of kinetic energy harvesters can increase their operational frequency range.

To compare the performance of a single generator with a wide bandwidth, a generator array and a single tunable generator with constant damping, typical specifications of these three types of generators have been chosen. G1, G2 and G3, listed in Table 1.7, represent a single generator with a wide bandwidth, a generator array and a single tunable generator with constant damping, respectively. Figure 1.80 shows the comparison of power spectra of these three types of generator.

Table 1.7 List of specifications in Fig. 1.80

Figure	1.80a	1.80b	1.80c	1.80d
Operational frequency range	90.5–110.5%	90.5–110.5%	61–160.5%	61–160.5%
Q-factor of G1	5	5	1	1
Q-factor of G2	50	80	10	50
Number of individual generator in G2, n	14	20	33	100
Resonant frequencies of individual generators in G2	$90\%+n\times1.4\%$	$90\%+n\times1\%$	$60\%+n\times3\%$	$60\%+n\times1\%$
Q-factor of G3	110	110	160	160
Mass ratio	40:2:1	40:1.3:1	200:4:1	200:1.3:1

In Fig. 1.80a, curves 1 and 3 are both single generators. Curve 2 consists of a generator array of 14 generators of Q-factor of 50. In Fig. 1.80b, curves 1 and 3 are identical to Fig. 1.80a. Curve 2 now consists of a generator array of 20 generators of Q-factor of 80. In Fig. 1.80c, for curve 1, the Q-factor has been reduced to 1. Curve 2 shows a generator array of 33 generators of Q-factor of 10. Curve 3 has the same Q-factor as in Fig. 1.80a and b but with higher mass. In Fig. 1.80d, curves 1

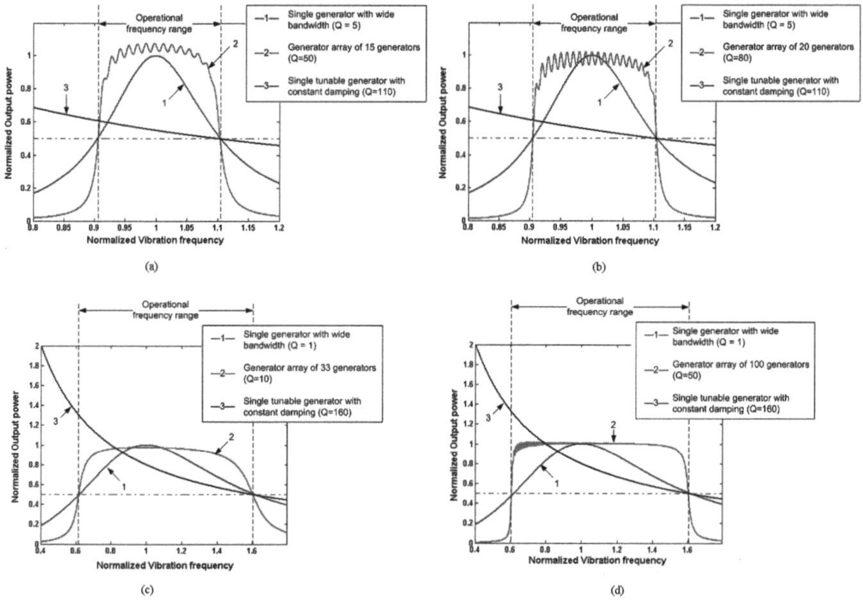

Fig. 1.80 Comparison of a single generator with a wide bandwidth, a generator array and a single tunable generator with constant damping

and 3 are identical to Fig. 1.80c. Curve 2 now consists of a generator array of 100 generators of Q-factor of 50.

When the Q-factor of a single generator decreases, its bandwidth increases. To generate the same amount of output power as the original bandwidth, a single generator has to be larger as the bandwidth increases.

If a generator array is used to widen the operational frequency range, one can design a few larger individual generators with low Q-factor with large resonant frequency gap between generators or many smaller individual generators with high Q-factor but small resonant frequency gap between generators. By contrast, it is much easier to design small tunable generators with constant damping to cover the same amount of operational frequency range at the cost of extra energy to power the frequency-tuning mechanisms.

Table 1.8 compares the advantages and disadvantages of different strategies to realize adaptive kinetic energy harvesters.

1.8 Summary

Several power supply strategies for wireless sensor networks have been introduced in this chapter. Attention has been paid especially to vibration-based microgenerators. A particular transduction mechanism is used to extract electrical energy

Table 1.8 Comparisons of different strategies

Strategies	Advantages	Disadvantages
Mechanical tuning	• High efficiency	o Extra system and energy are required o Responds to only one frequency at a time o Slow response to change in vibration frequency
◇ Change dimension	• Does not affect damping	o Difficult to implement o Not suitable for tuning in situ[a]
◇ Change centre of gravity	• Does not affect damping	o Not suitable for tuning in situ
◇ Change spring stiffness continuously	• Suitable for in situ tuning	o Consumes energy when generators work at resonance
◇ Apply axial load (change spring stiffness intermittently)	• Easy to implement • Suitable for in situ tuning • No energy is required when generators work at resonance • Damping is not affected when tensile load is applied	o Increased damping when compressive load is applied
Electrical tuning	• Easy to implement • No energy is required when generators work at resonance • Suitable for in situ tuning	o Low tuning efficiency o Complexity in designing the generator
Widen bandwidth	• No tuning mechanism required • Respond to different frequencies at the same time • Immediate response to change in vibration frequency	o Complexity in design
◇ Generator array	• Damping is not affected	o Complexity in design o Low volume efficiency
◇ Use mechanical stopper	• Easy to implement	o Fatigue problem o Decrease in maximum output power
◇ Coupled oscillators	• Easy to implement	o Decrease in maximum output power
◇ Non-linear generators	• Better performance at excitation frequencies higher than resonant frequency	o Complexity in design o Hysteresis
◇ Bi-stable structure	• Better performance at excitation frequencies much lower than resonant frequency	o Complexity in design

[a]In situ tuning: Tuning while the generator is mounted on the vibration source and working

from motion. The main transduction mechanisms are electromagnetic, electrostatic, piezoelectric and magnetostrictive.

Equation (1.21) gives a good guideline in designing kinetic energy harvesters. It is found that the maximum power converted from the mechanical domain to the electrical domain is proportional to the mass and vibration acceleration and inversely proportional to resonant frequency as well as mechanical (electrical) damping factor. This means that more power can be extracted if the inertial mass is increased or the generator can work in the environment where the vibration level is high. For a fixed resonant frequency, the generator has to be designed to make the mechanical damping as small as possible. For a generator with constant mechanical damping, the generated electrical power drops with an increase of the resonant frequency.

As most practical applications for kinetic energy harvesters exhibit frequency variations over time, it is not possible to guarantee that fixed frequency generators will always work at resonance and produce maximum output power. Mechanisms have to be employed to increase the operational frequency range of kinetic energy harvesters. Therefore, adaptive kinetic energy harvesters are developed. Generally, there are two possible solutions to adaptive kinetic energy harvesting. One is to tune the resonant frequency of a single generator and the other is to widen the bandwidth of the generator.

To tune the resonant frequency of a single generator, a certain mechanism has to be employed to periodically adjust the resonant frequency to match the frequency of ambient vibration at all times. Maximum power can then be generated at various frequencies without reducing the Q-factor and with high efficiency per unit volume. Tuning mechanisms can be classified as intermittent tuning and continuous tuning. Intermittent tuning has advantages over continuous tuning; in intermittent tuning, mechanism is turned off when the generator operates at the desired frequency thereby consuming negligible energy, which makes producing a net output power more probable. There are two methods to tune the resonant frequency: mechanical tuning and electrical tuning.

Among mechanical methods of frequency tuning, changing the dimensions of the structure and the position of the centre of gravity are potentially suitable for intermittent tuning. However, it is problematic to change and maintain the new dimensions of the structure or the centre of gravity of the proof mass during operation. The most suitable approach to changing the dimensions of the structure is to change its length. This requires that the structure clamp is removed, the length adjusted and then the structure re-clamped. If the structure cannot be clamped properly after tuning finishes, the performance of the generator will be severely affected by introducing damping effects through the supports. To change the position of the centre of gravity of the mass during operation, an actuator has to be embedded in the mass, which increases the complexity of the device. Therefore, these two methods are not suitable for in situ tuning (tuning while the generator is mounted on the vibration source and working) or tuning with automatic control.

Alternatively the frequency can be tuned by changing the spring stiffness intermittently or continuously. They are both suitable for in situ tuning but intermittently changing the spring stiffness is always preferred for efficiency reasons. However, extra systems and energy are required to realize tuning using mechanical methods.

It is important to mention that the efficiency of mechanical tuning methods also depends on the size of the structure. The smaller the resonator, the higher the efficiency of the tuning mechanism.

Resonant frequency tuning by adjusting the electrical load has been practically shown to be feasible. This method consumes little energy as it does not involve any change in mechanical properties. Energy is only consumed in electronic switches and control unit, which is typically far less than that consumed in mechanical tuning methods. In addition, it is much easier to be implemented than mechanical methods. However, the tuning efficiency of electrical tuning method to date is quite low and this method cannot achieve a large tuning range. It is a more suitable method when tunable frequency range required is small. An extra closed-loop system also has to be introduced to control the tuning process.

It is concluded that when choosing frequency method for a certain application, following factors need to be taken into consideration:

- The energy consumed by the tuning mechanism should be as small as possible and must not exceed the energy produced by the generator.
- The mechanism should achieve a sufficient operational frequency range.
- The tuning mechanism should achieve a suitable degree of frequency resolution.
- The generator should have as high as possible Q-factor to achieve maximum power output and the strategy applied should not increase the damping, i.e. decrease Q-factor, over the entire operational frequency range.

For the second solution, i.e. to widen the bandwidth, there is a trade-off between the system bandwidth and the Q-factor. Wider bandwidth means, for a single resonator, a lower Q-factor, which reduces the maximum output power. Bandwidth can also be effectively widened by designing a generator consisting of an array of small generators, each of which works at a different frequency. Thus, the assembled generator has a wide operational frequency range while the Q-factor does not decrease. However, this assembled generator must be carefully designed so that each individual generator does not affect the others. This makes it more complex to design and fabricate. Additionally, at a particular vibration frequency, only a single or a few individual generators contribute to power output so the approach is volume inefficient.

Another method used to increase the bandwidth is to use an amplitude limiter to limit the amplitude of the resonator. The drawbacks are that this method causes the maximum output power to drop by limiting vibration amplitude and the repeating mechanical contact between the cantilever and the mechanical stopper may result in earlier fatigue-induced failure in the cantilever beam.

Employing a coupled oscillator can also increase the operational bandwidth of the generator. It can achieve flat response over a wide frequency range. However, the maximum output power of a coupled oscillator generator is significantly lower than a generator with a single mass.

Furthermore, non-linear generators and generators with bi-stable structures are two further potential solutions to increase the operational frequency range of kinetic energy harvesters. They can improve performance of the generator at higher and lower frequency bands relative to its resonant frequency, respectively. However, the mathematical modelling of these generators is more complicated than that of linear generators, which increases the complexity in design and implementation. Besides, there is hysteresis in non-linear generators. Performance during down-sweep (or up-sweep) can be worse than that during up-sweep (or down-sweep) or worse than the linear region depending on sweep direction as explained in Section 1.6.

In conclusion, for vibration energy harvesting, possible strategies to increase the operation frequency range include

- changing spring stiffness intermittently (preferred) or continuously;
- straining the structure intermittently (preferred) or continuously;
- adjusting capacitive load;
- using generator array; and
- employing non-linear and bi-stable structures.

To realize these strategies properly, the following issues have to be considered. For intermittent mechanical tuning, the tuning system has to be designed to consume as little energy as possible and not to affect the damping so as to make the generator harvest maximum power. In addition, currently commercially available linear actuators are still large in size compared to millimetre-scale micro-generator. To keep tunable generators of reasonable size, it is important to use miniature actuators. Generators capable of electrical tuning must have strong electromechanical coupling to enable larger tuning ranges. Moreover, theoretical analyses of non-linear generators and generators with bi-stable structures have not been sufficiently developed and further attention should be paid to practically implement them.

Kinetic energy harvesting has been well studied in the past decade. It has been regarded as one of the best alternatives to energy source for wireless sensor networks. However, its drawback of narrow operational frequency bandwidth severely limits its application. With the recent development of adaptive kinetic energy harvesting, this drawback will eventually overcome, which will bring kinetic energy harvesting to much broader applications.

References

1. Roundy S, Wright PK, Rabaey J (2003) A study of low level vibrations as a power source for wireless sensor nodes. Comp. Comm. 26: 1131–1144
2. Beeby SP, Tudor MJ, White NM (2006) Energy harvesting vibration sources for microsystems applications. Meas. Sci. Tech. 17: 175–195
3. Zhu D, Tudor MJ, Beeby SP (2010) Strategies for increasing the operating frequency range of vibration energy harvesters: a review. Meas. Sci. Tech. 21(2): 022001
4. Williams CB, Yates RB (1996) Analysis of a micro-electric generator for microsystems. Sens. Actuator A 52: 8–11

5. Arnold DP (2007) Review of microscale magnetic power generation. IEEE Tran. Mag. 43(11): 3940–3951
6. Roundy S, Wright P, Pister K (2002) Micro-electrostatic vibration-to-electricity converters. Proceedings of IMECE, 1–10
7. Sodano HA, Inman DJ, Park G (2004) A review of power harvesting from vibration using piezoelectric materials. Shoock Vibr. Dig. 36(3): 197–205
8. Beeby SP, Tudor MJ, Torah RN, Roberts S, ODonnell T, Roy S (2007) Experimental comparison of macro and micro scale electromagnetic vibration powered generators. Microsyst. Tech. 13: 1647–1653
9. Stephen NG (2006) On energy harvesting from ambient vibration. J. Sound Vibr. 293: 409–25
10. Williams CB, Shearwood C, Harradine MA, Mellor PH, Birch TS, Yates RB (2001) Development of an electromagnetic micro-generator. IEE Proc. Circuits Dev. Syst. 148: 337–342
11. Ching NNH, Wong HY, Li WJ, Leong PHW, Wen Z (2002) A laser-micromachined vibrational to electrical power transducer for wireless sensing systems. Sens. Actuator A 97–98: 685–690
12. Glynne-Jones P, Tudor MJ, Beeby SP, White NM (2004) An electromagnetic, vibration-powered generator for intelligent sensor systems. Sens. Actuator A 110: 344–349
13. Koukarenko E, Beeby SP, Tudor MJ, White NM, ODonnell T, Saha T, Kulkani S, Roy S (2006) Microelectromechanical systems vibration powered electromagnetic generator for wireless sensor applications. Microsys. Tech. 12 (11): 1071–1077
14. Saha CR, ODonnell T, Loder H, Beeby SP, Tudor MJ (2006) Optimization of an electromagnetic energy harvesting device. IEEE Trans. Magnetics 42(10): 3509–3511
15. Beeby SP, Torah RN, Tudor MJ, Glynne-Jones P, ODonnell T, Saha CR, Roy S (2006) A micro electromagnetic generator for vibration energy harvesting. J. Micromech. Microeng. 17: 1257–65
16. Klahand H, Najafi K (2008) Energy scavenging from low-frequency vibrations by using frequency up-conversion for wireless sensor applications. IEEE Sensors J. 8(3): 261–268
17. Torah RN, Glynne-Jones P, Tudor MJ, ODonnell T, Roy S, Beeby SP (2008) Self-powered autonomous wireless sensor node using vibration energy harvesting. Meas. Sci. Tech. 19: 125202
18. Wang P, Dai X, Zhao X, Ding G (2009) A micro electromagnetic vibration energy harvester with sandwiched structure and air channel for high energy conversion efficiency. Proceedings of PowerMEMS 2009, Washington, DC, 296–299
19. Anton SR, Sodano HA (2007) A review of power harvesting using piezoelectric materials (2003–2006). Smart Mater. Struct. 16: 1–21
20. Sodano HA, Inman DJ (2005) Comparison of piezoelectric energy harvesting devices for recharging batteries. J. Intell. Mater. Syst. Struct. 16(10): 799–807
21. Roundy S, Wright PH, Rabaey JM (2004) Energy Scavenging for Wireless Sensor Networks with Special Focus on Vibrations. Kluwer, Norwell, MAs
22. MEMSnet http://www.memsnet.org/material.Cited18May2010
23. White NM, Glynne-Jones P, Beeby SP (2001) A novel thick-film piezoelectric micro-generator. Smart Mater. Struct. 10: 850–852
24. Lu F, Lee HP, Lim SP (2004) Modeling and analysis of micro piezoelectric power generators for micro-electromechanical-systems applications. Smart Mater. Struct. 13: 57–63
25. Jeon YB, Sood R, Jeong J-h, Kim SG (2005) MEMS power generator with transverse mode thin film PZT. Sens. Actuator A 122: 16–22
26. Fang HB, Liu J-Q, Xu ZY, Dong L, Wang L, Chen D, Cai BC, Liu Y (2006) Fabrication and performance of MEMS-based piezoelectric power generator for vibration energy harvesting. Microelectronics J. 37: 1280–1284
27. Marzencki M, Ammar Y, Basrour S (2007) Integrated power harvesting system including a MEMS generator and a power management circuit. Sens. Actuator. A 145–146: 363–370
28. Jeong S-J, Kim M-S, Song J-S, Lee H-K (2008) Two-layered piezoelectric bender device for micro-power generator. Sens. Actuator A 148: 158–167

29. Kok SW, White NW, Harris NH (2008) A novel piezoelectric thick-film free-standing cantilever energy harvester. EUROSENSORS XXII, Dresden, Germany, 395–399
30. Shen D, Park J-H, Ajitsaria J, Choe S-Y, Wikle III HC, Kim DJ (2008) The design, fabrication and evaluation of a MEMS PZT cantilever with an integrated Si proof mass for vibration energy harvesting. J. Micromech. Microeng. 18: 055017
31. Zhu D, Beeby SP, Tudor MJ, Harris NR (2009) A self powered smart tag for wireless structure health monitoring in aeronautical applications. Proceedings of PowerMEMS 2009, Washington, DC, 201–204
32. Meninger S (1999) A low power controller for a MEMS based energy converter. MSc thesis, MIT, USA
33. Mitcheson PD, Green TC, Yeatman EM, Holmes AS (2004) Architectures for vibration-driven micropower generators., IEEE J. Microelectromech. Syst. 13: 429–440
34. Meninger S, Mur-Miranda J, Lang J, Chandrakasan A, Slocum A, Schmidt M, Amirtharajah R (2001) Vibration to electric energy conversion. IEEE Trans. VLSI Syst. 9: 64–76
35. Tashiro R, Kabei N, Katayama K, Tsuboi F, Tsuchiya K (2002) Development of a electrostatic generator for a cardiac pacemaker that harnesses the ventricular wall motion. J. Artif. Organs 5(4): 239–45
36. Mitcheson P, Stark B, Miao P, Yeatman E, Holmes A, Green T (2003) Analysis and optimisation of MEMS on-chip power supply for self powering of slow moving sensors. Eurosensors 03, 17th European conference on sensors and actuators, University of Minho, Guimaraes, Portugal, 48–51
37. Arakawa Y, Suzuki Y, Kasagi N (2004) Micro seismic power generator using electret polymer film. Proceedings of PowerMEMS 2004, Kyoto, Japan, 187–190
38. Despesse G, Jager T, Chaillout J, Leger J, Vassilev A, Basrour S, Chalot B (2005) Fabrication and characterisation of high damping electrostatic micro devices for vibration energy scavenging. Proceedings of Design, Test, Integration and Packaging of MEMS and MOEMS, 386–390
39. Kuehne I, Frey A, Eckstein G, Schmid U, Seidel Haging (2006) Design and analysis of a capacitive vibration-to-electrical energy converter with built-in voltage. Proceedings of the 36th European Solid-State Device Research Conference 138–141
40. Yen BC, Lang JH (2006) A variable-capacitance vibration-to-electric energy harvester. IEEE Trans. Circuits Syst.-I: Regular Papers 53(2): 288–295
41. Sterken T, Fiorinil P, Altena G, Van Hoof C, Puers R (2007) Harvesting energy from vibrations by a micromachined electret generator. International Solid-State Sensors, Actuators and Microsystems Conference, Lyon, France, 129–132
42. Lo H, Tai YC (2008) Parylene-based electret power generators. J. Micromech. Microeng. 18: 104006
43. Hoffmann D, Folkmer B, Manoli Y (2009) Fabrication, characterization and modelling of electrostatic micro-generators. J. Micromech. Microeng. 19: 094001
44. Naruse Y, Matsubara N, Mabuchi K, Izumi M, Suzuki S (2009) Electrostatic micro power generation from low-frequency vibration such as human motion. J. Micromech. Microeng. 19: 094002
45. Huang J, O'Handley RC, Bono D (2003) New, high-sensitivity, hybrid magnetostrictive/electroactive magnetic field sensors. Proceedings of the SPIE 5050. 229–237
46. Wang L, Yuan FG (2008) Vibration energy harvesting by magnetostrictive material. Smart Mater. Struct. 17: 045009
47. Dai X, Wen Y, Li P, Yang J, Zhang G (2009) Modeling, characterization and fabrication of vibration energy harvester using Terfenol-D/PZT/Terfenol-D composite transducer. Sens. Actuator A 156: 350–358
48. Roundy S (2005) On the effectiveness of vibration-based energy harvesting. J. Intell. Mater. Syst. Struct. 16: 809–823
49. Mide Technology. http://www.mide.com/.Cited21May2010
50. Perpetuum Ltd. http://www.perpetuum.com/.Cited21May2010

51. Blevins RD (2001) Formulas for natural frequency and mode shape. Krieger, New York, NY
52. Gieras JF, Oh J-H, Huzmezan M, Sane HS (2007) Electromechanical energy harvesting system. Patent publication number: WO2007070022 (A2), WO2007070022 (A3)
53. Roylance L, Angell JB (1979) A batch fabricated silicon accelerometer. IEEE Trans. Electron Dev. 26(12): 1911–1917
54. Wu X, Lin J, Kato S, Zhang K, Ren T, Liu L (2008) A frequency adjustable vibration energy harvester. Proceedings of PowerMEMS 2008, Sendai, Japan, 245–248
55. Scheibner D, Mehner J, Reuter D, Kotarsky U, Gessner T, Dtzel W (2004) Characterization and self-test of electrostatically tunable resonators for frequency selective vibration measurements. Sen. Actuator. A 111: 93–99
56. Scheibner D, Mehner J, Reuter D, Gessner T, Dtzel W (2005) A spectral vibration detection system based on tunable micromechanical resonators. Sens. Actuator A 123–124: 63–72
57. Adams SG, Bertscht FM, Shawt KA, Hartwell PG, MacDonald NC, Moon FC (1995) Capacitance based tunable micromechanical resonators. The 8th International Conference on Solid-State Sensors and Actuators, and Eurosensors IX, Stockholm, Sweden, 438–441
58. Lee KB, Lin L, Cho YH (2008) A closed-form approach for frequency tunable comb resonators with curved finger contour. Sens. Actuator A 141: 523–529
59. Piazza G, Abdolvand R, Ho GK, Ayazi F (2004) Voltage-tunable piezoelectrically transduced single-crystal silicon micromechanical resonators. Sens. Actuator A 111: 71–78
60. Yao JJ, MacDonald NC (1996) A micromachined, single-crystal silicon, tunable resonator. J. Micromech. Microeng. 6: 257–264
61. Thiesen J, O'Brian GP (2006) Energy harvester with adjustable resonant frequency. Patent publication number: WO2006046937 (A1), EP1803170 (A1), US2008129147 (A1), EP1803170 (A0), CN101002343 (A)
62. Peters C, Maurath D, Schock W, Mezger F, Manoli Y (2009) A closed-loop wide-range tunable mechanical resonator for energy harvesting systems. J. Micromech. Microeng. 19: 094004
63. Wischke M, Masur M, Goldschmidtboeing F, Woias P (2010) Electromagnetic vibration harvester with piezoelectrically tunable resonance frequency. J. Micromech. Microeng. 20: 035025
64. Challa VR, Prasad MG, Shi Y, Fisher FT (2008) A vibration energy harvesting device with bidirectional resonance frequency tunability. Smart Mater. Struct. 17: 015035
65. Remtema T, Lin L (2001) Active frequency tuning for micro resonators by localized thermal stressing effects. Sens. Actuator A 91: 326–332
66. Syms RRA (1998) Electrothermal frequency tuning of folded and coupled vibrating micromechanical resonators. J. Microelectromechan. Syst. 7(2): 164–171
67. Cabuz C, Fukatsu K, Hashmoto H, Shoji S, Kurabayashi T, Minami K, Esashi M (1994) Fine frequency tuning in resonant sensors. Proceedings of IEEE Workshop on MEMS, Oiso, Japan, 245–250
68. Leland ES, Wright PK (2006) Resonance tuning of piezoelectric vibration energy scavenging generators using compressive axial preload. Smart Mater. Struct. 15: 1413–1420
69. Mukherjee R (2007) MEMS resonator using frequency tuning. Patent publication number: US20070214890A1
70. Hu Y, Xue H, Hu H (2007) A piezoelectric power harvester with adjustable frequency through axial preloads. Smart Mater. Struct. 16: 1961–1966
71. Eichhorn C, Goldschmidtboeing F, Woias P (2009) Bidirectional frequency tuning of a piezoelectric energy converter based on a cantilever beam. J. Micromech. Microeng. 19: 094006
72. Zhu D, Roberts S, Tudor MJ, Beeby SP (2008) Closed loop frequency tuning of a vibration-based micro-generator. Proceedings of PowerMEMS, Sendai, Japan, 229–232
73. Zhu D, Roberts S, Tudor MJ, Beeby SP (2009) Design and experimental characterisation of a tunable vibration-based electromagnetic micro-generator. Sens. Actuator A 158(2): 284–293
74. Wu W-J, Chen Y-Y, Lee B-S, He J-J, Peng Y-T (2006) Tunable resonant frequency power harvesting devices. Proc. of SPIE 6169: 55–62

75. Charnegie D (2007) Frequency tuning concepts for piezoelectric cantilever beams and plates for energy harvesting. MSc Dissertation School of Engineering, University of Pittsburgh, Pittsburgh, PA
76. Cammarano A, Burrow SG, Barton DAW, Carrella A, Clare LR (2010) Tuning a resonant energy harvester using a generalized electrical load. Smart Mater. Struct. 19: 055003
77. Xue H, Hu Y, Wang Q-M (2008) Broadband piezoelectric energy harvesting devices using multiple bimorphs with different operating frequencies. IEEE Trans. Ultrason. Ferroelectr. Freq. Control 55(9): 2104–2108
78. Feng G-H, Hung J-C (2007) Optimal FOM designed piezoelectric microgenerator with energy harvesting in a wide vibration bandwidth. Proceedings of the 2nd IEEE International Conference on Nano/Micro Engineered and Molecular Systems, Bangkok, Thailand, 511–514
79. Ferrari M, Ferrari V, Guizzetti M, Marioli D, Taroni A (2008) Piezoelectric multifrequency energy converter for power harvesting in autonomous microsystems. Sens. Actuator A 142: 329–335
80. Sari I, Balkan T, Kulah H (2007) A wideband electromagnetic micro power generator for wireless microsystems International Solid-State. Sensors, Actuators and Microsystems Conference, Lyon, France, 275–278
81. Lin SC, Lee BS, Wu WJ, Lee CK (2009) Multi-cantilever piezoelectric MEMS generator in energy harvesting. IEEE Int. Ultrason. Symp. Proc. 755–758
82. Soliman MSM, Abdel-Rahman EM, El-Saadany EE, Mansour RR (2008) A wideband vibration-based energy harvester. J. Micromech. Microeng. 18: 115021
83. Soliman MSM, Abdel-Rahman EM, El-Saadany EE, Mansour RR (2009) A design procedure for wideband micropower generators. J. MEMS. 18(6): 1288–1299
84. Petropoulos T, Yeatman EM, Mitcheson PD (2004) MEMS coupled resonators for power generation and sensing. Micromechanics Europe, Leuven, Belgium
85. Ramlan R, Brennan MJ, Mace BR, Kovacic I (2010) Potential benefits of a non-linear stiffness in an energy harvesting device. Nonlin. Dyn. 59: 545–558
86. Spreemann D, Folkmer B, Maurath D, Manoli Y (2006) Tunable transducer for low frequency vibrational energy scavenging. Proceedings of EurosensorsXX, Göteborg, Sweden
87. Mann BP, Sims ND (2009) Energy harvesting from the nonlinear oscillations of magnetic levitation. J. Sound Vibr. 319. 515–530
88. Burrow SG, Clare LR (2007) A resonant generator with non-linear compliance for energy harvesting in high vibrational environments. IEMDC '07 1: 715–720
89. Burrow SG, Clare LR, Carrella A, Barton D (2008) Vibration energy harvesters with non-linear compliance. Proc. of SPIE. 6928: 692807
90. Tvedt LGW, Nguyen DS, Halvorsen E (2010) Nonlinear behavior of an electrostatic energy harvester under wide- and narrowband excitation. J. MEMS 19(2): 305–316
91. Galchev T, Kim H, Najafi K (2009) A parametric frequency increased power generator for scavenging low frequency ambient vibrations. Eurosensors XXIII Conference Procedia Chemistry 1: 1439–1442
92. Mann BP, Owens BA (2010) Investigations of a nonlinear energy harvester with a bistable potential well. J. Sound and Vibr. 329: 1215–1226
93. Ferrari M, Ferrari V, Guizzetti M, Ando B, Baglio S, Trigona C (2009) Improved energy harvesting from wideband vibrations by nonlinear piezoelectric Converters. Eurosensors XXIII Conference Procedia Chemistry 1: 1203–1206
94. Stanton SC, McGehee CC, Mann BP (2010) Nonlinear dynamics for broadband energy harvesting: Investigation of a bistable piezoelectric inertial generator. Physica D 239: 640–653

Chapter 2
Modelling, Performance Optimisation and Automated Design of Mixed-Technology Energy Harvester Systems

Tom J. Kaźmierski and Leran Wang

Abstract This chapter presents an automated energy harvester design flow which is based on a single HDL software platform that can be used to model, simulate, configure and optimise the complete mixed physical-domain energy harvester system (micro-generator, voltage booster, storage element and load). We developed an accurate HDL model for the energy harvester and demonstrated its accuracy by validating it experimentally and comparing it with recently reported models. A demonstrator prototype incorporating an electromagnetic mechanical-vibration-based micro-generator and a limited number of library models has been developed and a design case study has been carried out. Experimental measurements have validated the simulation results which show that the outcome from the design flow can improve the energy harvesting efficiency by 75%.

Keywords Kinetic energy harvester · Design flow · Optimisation · Mixed-domain modelling

2.1 Introduction

At present there are considerable and continuing research efforts worldwide to support the energy harvesting paradigm and self-powered electronics. The majority of the reported research in energy harvesting has been on improving the efficiency of the energy harvesters through the design and fabrication of novel micro-generators, materials and devices [3]. The amount of power that can be harvested in a particular application is highly dependent upon the energy source being harvested. Typically, power densities of around 800 $\mu W/cm^3$ for machine vibration applications and up to 140 $\mu W/cm^3$ for human-powered applications can be expected [4]; however, the power output of vibration-harvesting inertial generators is highly sensitive to the frequency and amplitude of the vibration source [8] and so these figures are indicative only. Practical generators have been reported with power densities of 17

T.J. Kaźmierski (✉)
School of Electronics and Computer Science, University of Southampton, Southampton, UK
e-mail: tjk@ecs.soton.ac.uk

T.J. Kaźmierski, S. Beeby (eds.), *Energy Harvesting Systems*,
DOI 10.1007/978-1-4419-7566-9_2, © Springer Science+Business Media, LLC 2011

$\mu W/cm^3$ for a non-resonant device [6, 9], to a resonant device capable of generating 30 $\mu W/cm^3$. Typically the generated voltage from a micro-generator is insufficient to power an electronic device directly, and therefore external analogue circuits are often employed to rectify and boost the voltage and store the energy in a battery or a supercapacitor. Various circuit designs have also been reported, such as an AC/DC rectifier combining with a switch-mode DC/DC converter [11] and self-timed circuits which eliminate AC/DC conversion [15]. Because of the output power of energy harvesters is highly variable and unpredictable, recently Amirtharajah et al. proposed digital circuits that employ serial computation and distributed arithmetic as a way of dealing with the variability of the available energy [1]. Also, in the publication of Wang et al., it was reported through the use of dedicated sub-threshold logic it is possible to implement an FFT processor that operates in practice with supply voltage as low as 180 mV, suitable for energy harvesters [20]. Electrical power management technique has also been investigated. "The charge-based control unit will make sure the energy available is enough for the atomic operation before triggering the computation" [14, 19].

An energy harvester has normally three main components: the micro-generator which converts ambient environment energy into electrical energy, the voltage booster which pumps up and regulates the generated voltage and the storage element (Fig. 2.1).

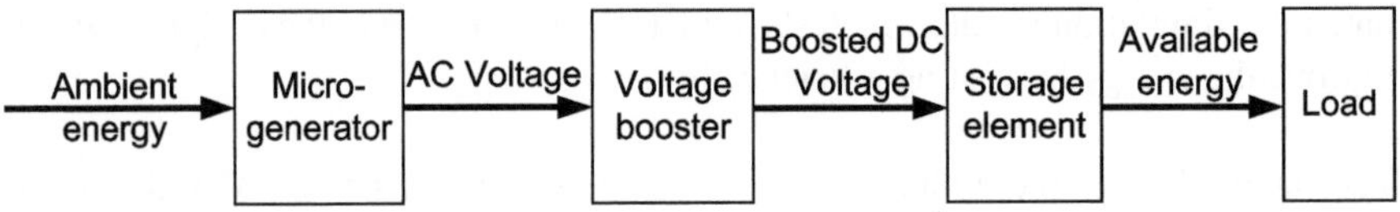

Fig. 2.1 Block diagram of an energy harvester

Clearly such an energy harvester consists of components from both mechanical and electrical domains as well as external circuits which regulate and store the generated energy. Therefore, the performance optimisation should only be based on a model that describes the energy harvester as an integrated system. However, most existing modelling and optimisation methods are concentrating on either the micro-generator [21] or the external circuits [1] separately while the design tools for an integrated system are missing. MATLAB and finite element analysis (FEA) packages are being used to simulate and optimise the performance of the micro-generator part of the self-powered system [2]. To design and optimise the energy harvester associated electronics, simulators such as SPICE are often used. The micro-generator is usually modelled either as an ideal voltage source [22] or an equivalent circuit model [1] because "the current EDA tools do not support direct integration of the electromechanical dynamics of vibration-based energy harvesters into circuit simulations" [1].

To design highly efficient energy harvesters, it is crucial to consider the various parts of an energy harvester in the context of a complete system, or the gain at one part may come at the price of efficiency loss elsewhere, rending the energy harvester much less efficient than expected. To date there has been no reported design flow for energy harvesters and the aim of this chapter is to propose such an automated design flow for the modelling, configuration, and optimisation of energy harvester systems through the use of VHDL-AMS. VHDL-AMS can be used to model and correctly predict the performance of an energy harvester system because it describes the micro-generator and external electronics as an integrated model, so that the close mechanical–electrical interaction, which is often missing in traditional energy harvester design methods, can be captured accurately.

2.2 Energy Harvester Design Flow

This chapter proposes an automated energy harvester design flow which is based on a single HDL software platform that can be used to model, simulate, configure and optimise energy harvester systems. The proposed design flow is outlined in the pseudo-code of Algorithm 1 and also shown in Fig. 2.2. Naturally, the process starts with initial design specification, such as available energy source (light, heat, vibration, etc.), environmental energy density, device size, minimum voltage level/power output. According to these specifications, HDL models are constructed from component cells available in the component library. The component library contains parameterised models of different kinds of micro-generator structures (solar cell, electromagnetic, piezoelectric, etc.), various booster circuit topologies and storage elements. The outer loop in the algorithm represents this structure configuration process, which involves examining and comparing those HDL models from the library with the aim of identifying a set of components that meet specific user requirements. The inner design flow loop will then find the best performance of each candidate design by adjusting electrical and non-electrical parameters of the design's mixed technology HDL model. The parametric optimisation of the generated structure will further improve the energy harvester efficiency by employing suitable optimisation algorithms. The design flow ends at the best performing design for fabrication subject to the user-defined performance characteristics.

Algorithm 1 Automated energy harvester design flow.

Initial design structure and specification
Structure configuration loop:
for all design structures **do**
  Build HDL model of design
  Optimisation loop:
  repeat
    Simulate and evaluate performance
    if best performance not achieved **then**
      Update design parameters
    end if
  until best performance achieved
  if there are more structures to try **then**
    Select new structure
  end if
end for

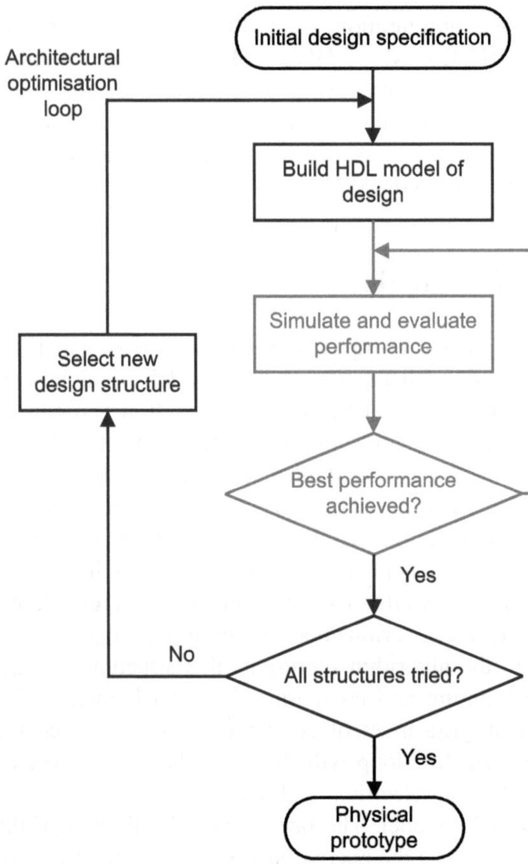

Fig. 2.2 Energy harvester design flow

The requirements for energy harvester component models are: (1) models need to be computationally efficient for fast performance optimisation when used in complete energy harvester systems and yet accurate, these are conflicting requirements; (2) models need to capture both theoretical equations and practical non-idealities required for accurate performance estimation. The models should support different mechanical–electrical structures and will be expressed in terms of HDL descriptions. They will be able to predict the behaviour of the actual device accurately while remaining reconfigurable.

As proof of concept, a small VHDL-AMS model library has been built to demonstrate the efficiency of the design flow, which is shown in Section 2.3. Based on the developed model library, Section 2.4 describes the automated structure configuration that has been carried out using a single VHDL-AMS simulator. The configuration result and simulations of different energy harvester models have led to in-depth understanding about how electromagnetic micro-generator performs when connected with voltage multipliers. The results have been used for the performance optimisation, which is presented in Section 2.5.

2.3 Energy Harvester Modelling

2.3.1 Micro-generator

Reported modelling approaches aim to replace the micro-generator of an energy harvester with either an ideal voltage source (Fig. 2.3a) [22] or an equivalent circuit model (Fig. 2.3b) [1] when designing the voltage booster. However, as will be shown later, neither of these approaches is suitable for accurate voltage booster design.

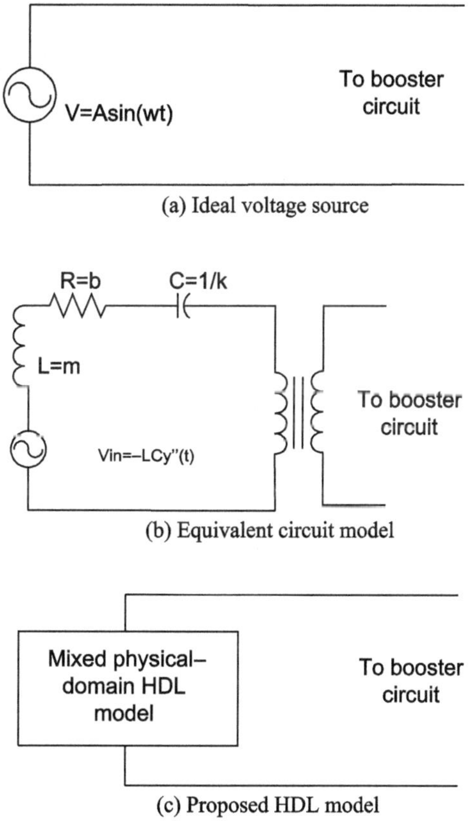

Fig. 2.3 Micro-generator models

The proposed approach uses VHDL-AMS to describe the micro-generator as a series of analytical equations (Fig. 2.3c), which includes mechanical, magnetic and electrical behaviours of the micro-generator. Throughout this section, comparisons have been made between different modelling approaches and it has been demonstrated that the proposed HDL-based model is more accurate than the circuit models.

The case study presented here uses a vibration-based electromagnetic micro-generator which was developed by Torah et al. [16] as an example. The design is based on a cantilever structure. The coil is fixed to the base and four magnets, which are located on both sides of the coil, form the proof mass (see Fig. 2.4).

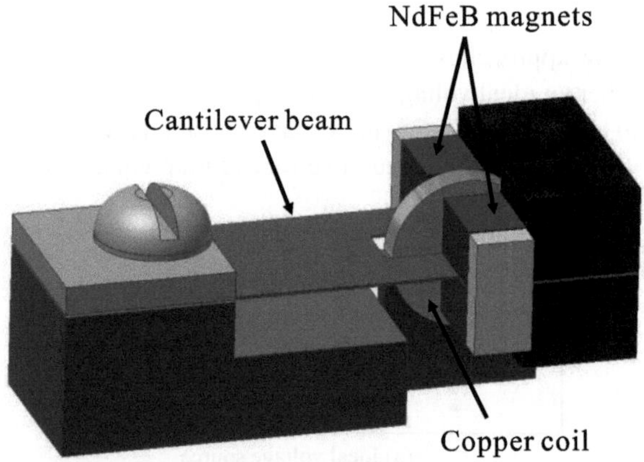

Fig. 2.4 Cantilever-based electromagnetic micro-generator [16]

This structure can be modelled as a second-order spring-damping system, which has been widely used, and whose dynamics is [3]

$$m\ddot{z}(t) + c_{\mathrm{p}}\dot{z}(t) + k_{\mathrm{s}}z(t) + F_{\mathrm{em}} = -m\ddot{y}(t) \tag{2.1}$$

where m is the proof mass, $z(t)$ is the relative displacement between the mass and the base, c_{p} is the parasitic damping factor, k_{s} is the spring stiffness, $y(t)$ is the displacement of the base and F_{em} is the electromagnetic force.

The electromagnetic voltage generated in the coil is given by Faraday's law:

$$v_{\mathrm{em}} = \Phi(z) * \dot{z}(t) \tag{2.2}$$

where $\Phi(z)$ is the magnetic flux through the coil.

Although the developed HDL model is based on analytical equations, it can capture practical size and shape of the actual device. The coil in the actual micro-generator consists of N turns and has an inner diameter r and outer diameter R. Each of the four opposite magnets are of height H (see Fig. 2.5a).

So the actual magnetic flux through the coil is a piecewise non-linear function of the relative displacement $z(t)$: $\Phi = f\{z(t)\}$.

When the relative displacement is small $|z(t)| < r$ (Fig. 2.5b):

$$\Phi = \left(\sqrt{R^2 - z^2(t)} + \sqrt{r^2 - z^2(t)}\right) * 2 * B * N \tag{2.3}$$

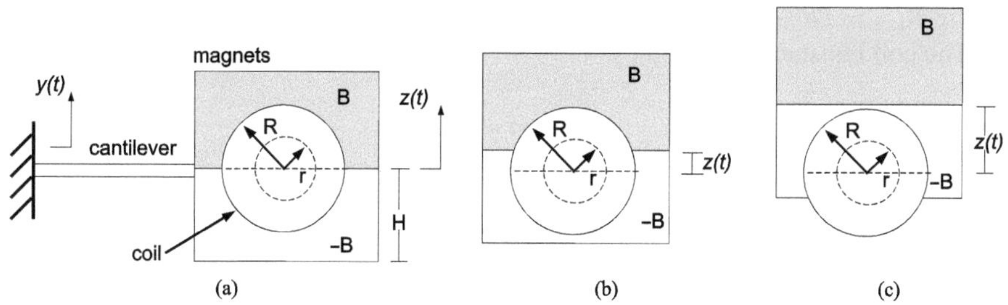

Fig. 2.5 Relative displacement between the coil and magnets in the micro-generator: (**a**) initial position $z(t) = 0$; (**b**) small displacement $|z(t)| < r$; and (**c**) large displacement $H - r < |z(t)| < H$

When the relative displacement is large $H - r < |z(t)| < H$ (Fig. 2.5c):

$$\Phi = -\left(\sqrt{R^2 - (H - |z(t)|)^2} + \sqrt{r^2 - (H - |z(t)|)^2}\right) * B * N \tag{2.4}$$

There are five other sections of the piecewise function which have been implemented in the VHDL-AMS model (see the code on page 80) but are omitted here.

The output voltage is simply defined by Kirchoff's voltage law:

$$v(t) = v_{\text{em}} - R_{\text{c}} * i(t) - L_{\text{c}} * \dot{i(t)} \tag{2.5}$$

where R_{c} and L_{c} are the resistance and inductance of the coil, respectively, and $i(t)$ is the current through the coil.

Finally, the electromagnetic force is calculated as (Lorentz force law):

$$F_{\text{em}} = \Phi(z) * i(t) \tag{2.6}$$

In the above equations, the coil parameters are given in number of turns N and resistance R_{c}. However, when manufacturing a coil, the specification is often given by the thickness t, inner radius r, outer radius R and wire diameter d. To build parameterised HDL models, the relations between these parameters are incorporated and listed below. These equations can be deducted using basic geometry and physics knowledge.

The total wire length is

$$l = \frac{4 * f * t * (R^2 - r^2)}{d^2} \tag{2.7}$$

where f is the fill factor.

The number of turns is

$$N = \frac{l}{2 * \pi * R_{\text{ave}}} \tag{2.8}$$

where $R_{\text{ave}} = (R - r)/2 + r$ is the average radius.

The coil resistance is given by

$$R_{\text{c}} = \frac{4 * \rho * l}{\pi * d^2} \tag{2.9}$$

where ρ is the resistivity of the material.

The VHDL-AMS code of the model is given below:

```
library IEEE;
use IEEE.ENERGY_SYSTEMS.all;
use IEEE.MECHANICAL_SYSTEMS.all;
use IEEE.ELECTRICAL_SYSTEMS.all;
use IEEE.math_real.all;
use work.EnergyHarvester.all;

entity EMH is
  port(terminal HOUSE:translational;
  terminal LOAD:electrical);
end entity EMH;

architecture Behaviour of EMH is
  quantity yt across HOUSE to translational_ref;
  quantity zt:DISPLACEMENT;
  quantity emv:VOLTAGE;
  quantity vt across it through LOAD to electrical_ref;
  quantity Fem,abszt,Pout,Phi:real;

begin
  mp*zt'DOT'DOT+Cp*zt'DOT+Ks*zt+Fem==-mp*yt'DOT'DOT;
  Phi*zt'DOT==emv;
  emv==vt-Rc*it-Lc*it'DOT;
  Fem==-Phi*it;
  Pout==-it*vt;
  abszt==abs(zt);
  if 0.0<=abszt and abszt<Rin use
  Phi==(sqrt(abs(Rout**2-zt**2))+sqrt(abs(Rin**2-zt**2)))*2.0*B*N;

  elsif Rin<=abszt and abszt<(Htm-Rout) use
  Phi==sqrt(abs(Rout**2-zt**2))*2.0*B*N*(Rout-abszt)/Rw;

  elsif (Htm-Rout)<=abszt and abszt<Rout use
  Phi==sqrt(abs(Rout**2-zt**2))*B*N*(Rout-abszt)/Rw+
  (sqrt(abs(Rout**2-zt**2))-sqrt(abs(Rout**2-(Htm-abszt)**2)))*B*N*
    (Rout-Htm+abszt)/Rw;

  elsif Rout<=abszt and abszt<(Htm-Rin) use
  Phi==-sqrt(abs(Rout**2-(Htm-abszt)**2))*B*N*(Rout-Htm+abszt)/Rw;

  elsif (Htm-Rin)<=abszt and abszt<Htm use
  Phi==-(sqrt(abs(Rout**2-(Htm-abszt)**2))
    +sqrt(abs(Rin**2-(Htm-abszt)**2)))*B*N;

  elsif Htm<=abszt and abszt<(Htm+Rin) use
  Phi==-(sqrt(abs(Rout**2-(abszt-Htm)**2))+
 sqrt(abs(Rin**2-(abszt-Htm)**2)))*B*N;
```

```
  elsif (Htm+Rin)<=abszt and abszt<(Htm+Rout) use
  Phi==-sqrt(abs(Rout**2-(abszt-Htm)**2))*B*N*(Rout-abszt+Htm)/Rw;

  else
  Phi==0.0;
  end use;
end architecture Behaviour;
```

It has been proved that the maximum average power P_{avelec} that can be delivered to the electrical domain is given by [2]

$$P_{\text{avelec}} = \frac{m^2 * Y^2 * \omega_n^4}{8 * c_{\text{p}}} \tag{2.10}$$

where Y is the amplitude of external kinetic excitation, $\omega_n = \sqrt{k_{\text{s}}/m}$ is the system's resonant frequency. Note that all the simulation and experimental tests in this chapter are based on a 50 Hz sine wave excitation with an amplitude of $Y = 8.4\,\mu\text{m}$.

This happens when the system's parasitic damping equals to the electromagnetic damping; therefore, if a load resistance R_{l} is connected to the micro-generator, its optimal value is [13]

$$R_{\text{loptimal}} = \Phi^2/c_{\text{p}} - R_{\text{c}} \tag{2.11}$$

Two types of this micro-generator have been modelled, which are based on the same structure but have different dimensions. Some of the key parameters are listed in Table 2.1. As can be seen from the table, micro-generator type II is bigger than type I and because the coil is changeable, both the micro-generators can have different wire diameters.

Table 2.1 Micro-generators' parameters

	Type I	Type II
Proof mass m (g)	0.6	2.4
Magnet height H (mm)	2.0	3.0
Field strength B (T)	0.5	0.7
Wire diameter d (μm)	12/16/25	16/25
Coil outer radius R (mm)	1.2	2.45
Coil thickness t (mm)	0.48	1.3

2.3.2 Voltage Booster

Voltage boosters are external circuits to the micro-generator that are used to boost up the output voltage and to perform necessary AC–DC rectification. There are a number of circuit topologies that can be used as a voltage booster. VHDL-AMS has also been used to describe the circuit behaviour by DAEs.

2.3.2.1 Voltage Multiplier

A voltage multiplier (VM), which uses cascaded diodes and capacitors to achieve higher DC voltage from an AC input, meets the requirements of a booster and has been investigated here. There are two types of voltage multiplier based on different configurations, namely Villard (Fig. 2.6a–d) and Dickson (Fig. 2.6e–h) [22]. In the model library, these VMs are configured as 3, 4, 5 and 6 stages.

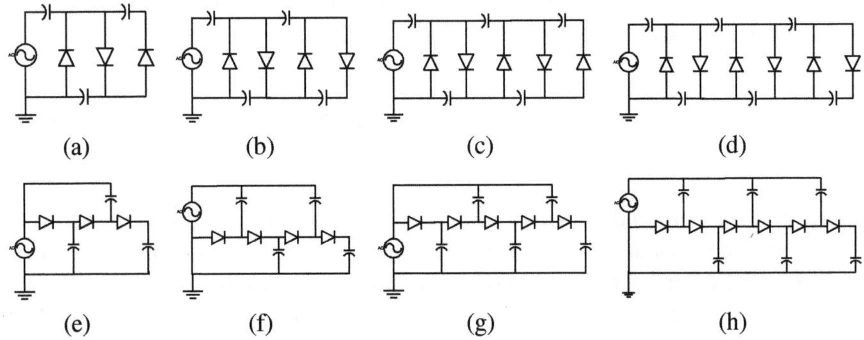

Fig. 2.6 Voltage multiplier configurations, **a–d**: three- to six-stage Villard VM, **e–h**: three- to six-stage Dickson VM

2.3.2.2 Voltage Transformer

A voltage transformer together with a full wave rectifier can also act as the voltage booster for an energy harvester. Two types of rectifier configurations have been tested. Simulation results show that comparing to a common full-bridge rectifier, the configuration in Fig. 2.7 gives better performance since it uses less diodes and thus loses less energy. The number of turns and the resistance value of primary ($N1$, $R1$) and secondary winding ($N2$, $R2$) are the four main parameters that determine the voltage transformer's performance.

2.3.3 Supercapacitor

In case of the storage element, a supercapacitor has been modelled as in Fig. 2.8 [10], where R_{leakage} represents the leakage resistance and R_{ESR} is the equivalent series resistance.

2.3.4 Models Comparison

This section compares the accuracy of different micro-generator modelling approaches with the same voltage booster. Type I micro-generator is connected to

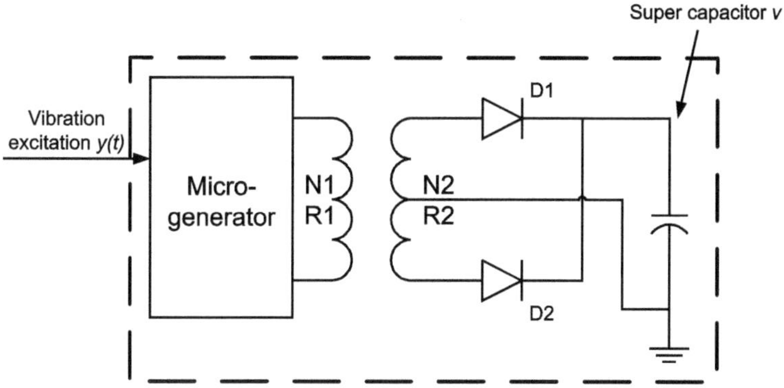

Fig. 2.7 Voltage transformer configuration

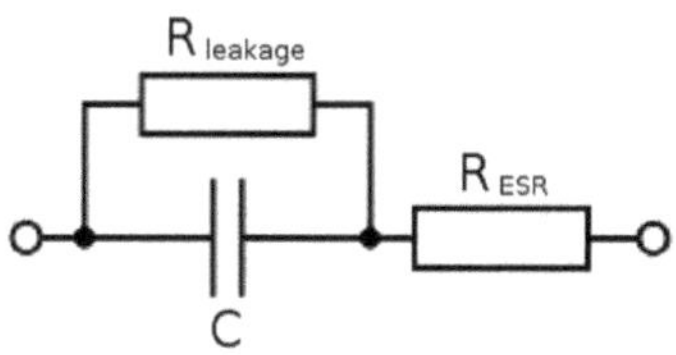

Fig. 2.8 Simplified supercapacitor model

a six-stage Villard voltage multiplier (Fig. 2.6d) and three models of this energy harvester system have been built.

It has been reported that when excited by a 50 Hz sine wave vibration of 8.4 μm amplitude, type I micro-generator can generate a maximum power of 45.7 μW under the optimal load condition and the output voltage is around 600 mV [16]. So an ideal voltage source (Fig. 2.3a) of 50 Hz frequency and 640 mV amplitude is connected to the VM circuit and SPICE circuit simulations have been carried out to evaluate the VM's performance.

The equivalent circuit model (Fig. 2.3b) of the micro-generator links mechanical mass (m), spring (k) and damper (b) to electrical inductor (L), capacitor (C) and resistor (R) by [1]

$$L = m, \quad C = 1/k, \quad R = b \tag{2.12}$$

and SPICE circuit simulations have also been carried out.

The VHDL-AMS model incorporates the micro-generator and the VM circuit booster.

The comparisons presented in this section are based on the charging of a 0.22 F supercapacitor. Because the VHDL-AMS simulator used here, SystemVision from Mentor Graphics [5], has a maximum simulation time of 150 min, only simulation results in this range have been obtained. Figure 2.9 shows the simulated charging waveform of the capacitor using different micro-generator models including the proposed HDL model. Also shown is the capacitor charging waveform obtained experimentally. The experiment was set up as the micro-generator sitting on a vibration generator, which produces constant mechanical vibrations (Fig. 2.10a), and the data are collected by LabView software (Fig. 2.10b).

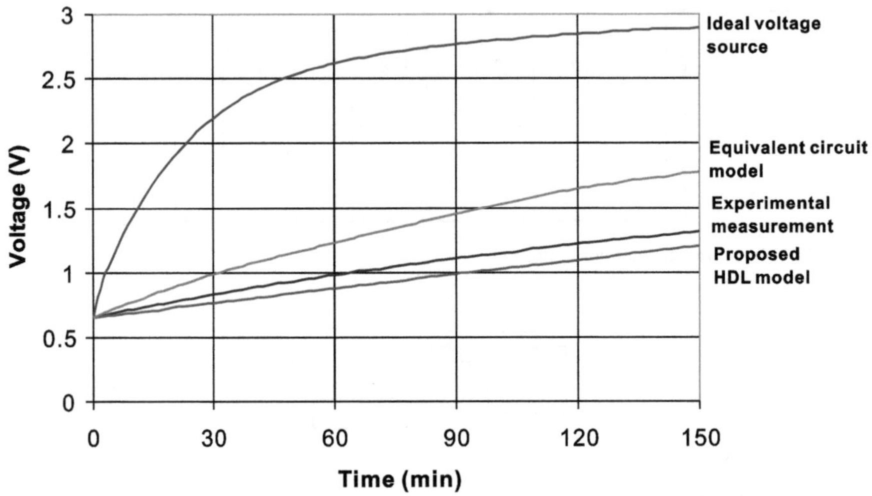

Fig. 2.9 Different energy harvester models compared with experimental measurements

As it can be seen, there is poor correlation between the energy harvester experimental measurements and that of models based on ideal voltage source or equivalent circuit. The reason why the model with ideal voltage source (Fig. 2.3a) fails is that in an energy harvester the voltage booster can greatly affect the behaviour of the micro-generator but an ideal voltage source always produces constant output. The crucial mechanical–electrical interaction is missing in the model.

As for the equivalent circuit model (Fig. 2.3b), this is because Eq. (2.12) is an over-simplification and does not capture accurately the internal operation of the energy harvester.

In the case of the proposed energy harvester HDL model, there is a good correlation between the experimental and simulation results. The reason why the HDL-based model correlates well with practice is that it can incorporate the actual shape and size of various components into the micro-generator model by using analytical equations. Here the non-linear dependence of the micro-generator's output voltage on the input displacement described in Section 2.3 can be accurately captured by the HDL model. Simulation results of the equivalent circuit model and HDL model are compared with the experimental measurement in Fig. 2.11. As can be seen from the waveforms, when excited by a sine wave stimulus, the equivalent circuit model still generates sine wave output. But the HDL model can capture the situations when the coil and magnets are moving apart, which is what happens in practice and leads to non-sine wave output.

(a)

(b)

Fig. 2.10 Experimental measurement set-up: (**a**) micro-generator sitting on a vibration generator and (**b**) Labview software collecting the data

Although the simulation results of the proposed HDL models are very close to the experimental measurements, there are still notable differences between them in both Figs. 2.9 and 2.11. The reasons may be found by examining the practical conditions during experiments. In Fig. 2.9, the charging waveforms do not start from 0 V because in practical measurements the capacitors all had a bit of initial charge and adjustments on the timing have been made to the simulation waveforms so that all the curves start at the same voltage (about 0.6 V), which may lead to the difference. In Fig. 2.11, the difference between the simulation waveform and experimental measurement may be generated by the interference that was brought in by the oscilloscope's probes.

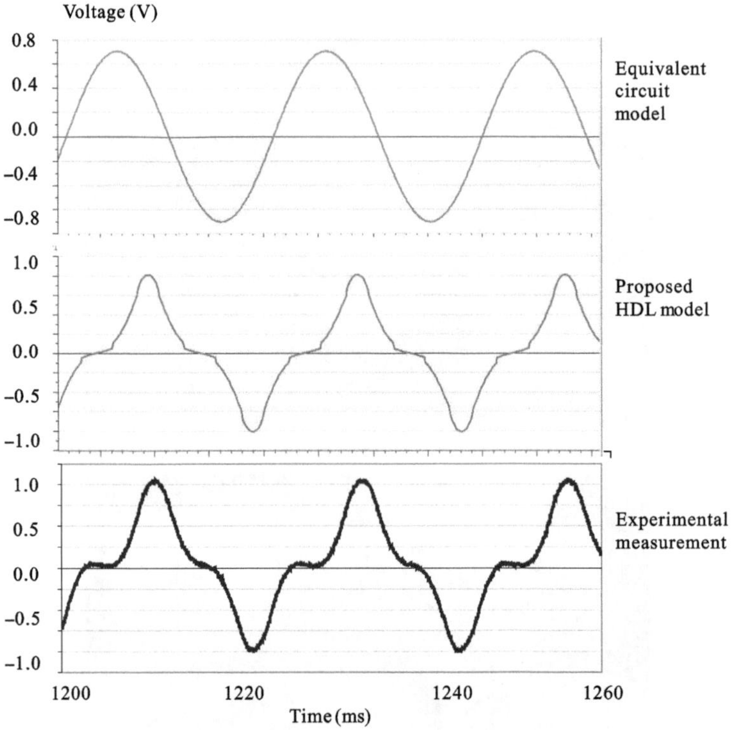

Fig. 2.11 Non-linear output from micro-generator correctly reflected by proposed model

2.4 Automatic Structure Configuration Based on HDL Model Library

From the last section, a small HDL model library of energy harvester components has been built. It contains two types of micro-generator, each of which can be configured with different coils (wire diameter of 12/16/25 μm) and two types of voltage multipliers that have three to six stages. The voltage transformer has not been included because it cannot be made and tested with available resources. But the simulation-based optimisation of energy harvester with voltage transformer has been performed and will be discussed in Section 2.5.2. The configuration target has been set to find the set of components that can charge the 0.047 F supercapacitor to 2 V in shortest time. These values were chosen because there has been reported energy harvester systems that use 0.047 F storage capacitor and 2 V working voltage [17].

Simulations of every available energy harvester configuration were carried out simultaneously and a process has been developed to automatically track the best model. SystemVision VHDL-AMS simulator [5] has been used as the single software platform. The outcome design is listed in Table 2.2.

Table 2.2 Parameters of the configuration result

Micro-generator	Type II
Wire diameter	25 μm
Voltage booster	Three-stage Dickson VM (Fig. 2.6e)

Not surprisingly, the micro-generator II has been chosen because it is bigger and stores more kinetic energy. However, it is quite interesting that the coil with the largest wire diameter, which leads to fewest number of turns, and the VM with fewest stages have been chosen. To further investigate on this result, more simulations have been done and an important trade-off between the electromagnetic micro-generator and the VM voltage booster has been found.

Figure 2.12 shows the charging waveforms of type I micro-generator connected to the same five-stage VM but configured with different coils. At the beginning, the energy harvester with 25 μm wire diameter charges the quickest and the 12 μm configuration charges the slowest while the 16 μm one is in between. But the 25 μm configuration also saturates quickly and reaches the 2 V mark slower than the 16 μm energy harvester. Due to simulation time limitation, the figure does not show how the other two waveforms end. But it could be foreseen that the 16 μm configuration will also saturate at some point while the 12 μm one reaches highest voltage.

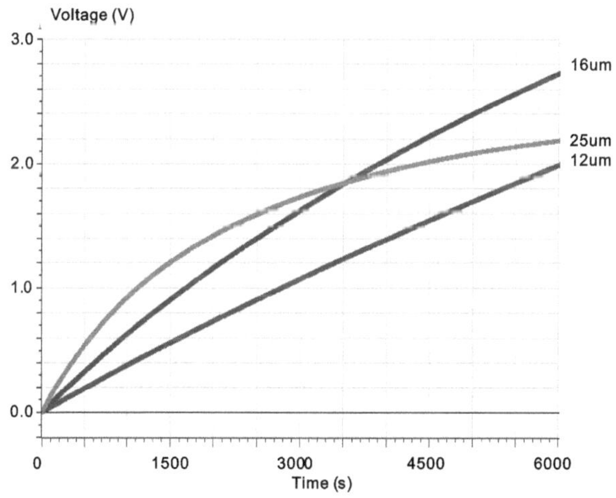

Fig. 2.12 Simulation of type I micro-generator with different coils

Similar results have been obtained from the voltage booster end. Figure 2.13 shows the charging waveforms of type II micro-generator with 25 μm coil connecting with three-, four- and five-stage Dickson VMs. It can be seen that the energy harvester with three-stage VM charges the supercapacitor to 2 V first and the one with five-stage VM can reach the highest voltage.

From the simulation results it can be concluded that in an energy harvester design that combines electromagnetic micro-generator and voltage multiplier, the fewer

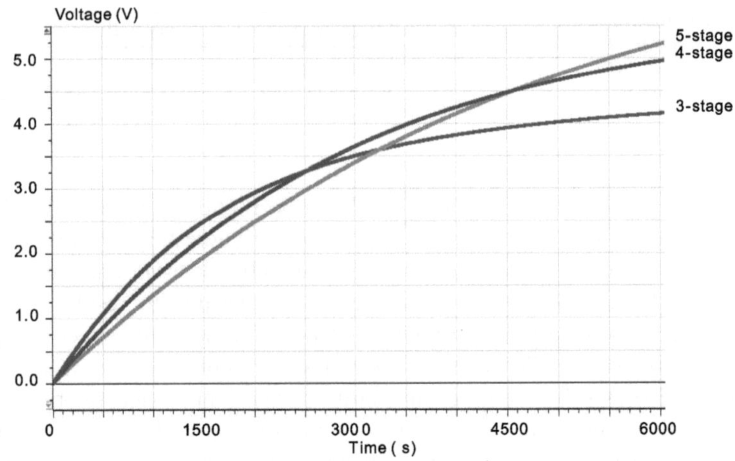

Fig. 2.13 Simulation of type II micro-generator with different VMs

number of turns in the coil and the fewer VM stages, the higher initial charging rate the energy harvester can get but the lower voltage it can finally reach. Therefore, although the micro-generator with more coil turns can generate more power and VMs with more stages can boost the voltage higher, under certain circumstances the optimisation of subsystems in isolation does not lead to a globally optimised design. It proves that when combining different components of an energy harvester, the gain at one part may come at the price of efficiency loss elsewhere, rending the energy harvester much less efficient than expected. This information is very useful for the development of future, more complicated systems and model libraries.

2.5 Performance Optimisation

The close mechanical–electrical interaction (micro-generator and voltage booster) that takes place in energy harvesters often leads to significant performance loss when the various parts of the energy harvesters are combined. Here the loss expressed in terms of energy harvesting efficiency

$$\eta_{\text{Loss}} = \frac{E_{\text{Harvested}} - E_{\text{Delivered}}}{E_{\text{Harvested}}} \tag{2.13}$$

In the proposed design flow, the generated energy harvester design should be parameterised such that automated performance optimisation will be able to further improve the energy harvester efficiency by employing suitable optimisation algorithms. The parameters used for the optimisation are from both the micro-generator and the voltage booster. The optimisation object is to increase the charging rate of the supercapacitor.

2.5.1 Exhaustive Search

The micro-generator parameters that can be optimised are related to the coil size, i.e. the thickness (t) and the outer radius (R), because other components such as the magnets and cantilever determine the resonant frequency of the micro-generator and thus should be based on application requirements. The circuit parameters of voltage booster are the capacitor values of each VM stage. The entire energy harvester is optimised as an integrated model. The searching space of parameters has been given in Table 2.3

Table 2.3 Optimisation searching space

Coil thickness (mm)	1.0–1.3
Coil radius (mm)	2.0–2.45
Capacitor values (μF)	47/100/150

The optimisation is based on the concurrent simulations of design instances from uniform sampling of the search space and tracking the best result (Fig. 2.14). This is relatively simple and straightforward because after the automatic structure configuration the search space is quite small and the VM capacitors can only have discrete values. However, other optimisation algorithms may also be employed and in Section 2.5.2 a VHDL-AMS-based genetic optimisation has been successfully applied to the integrated optimisation of energy harvester systems.

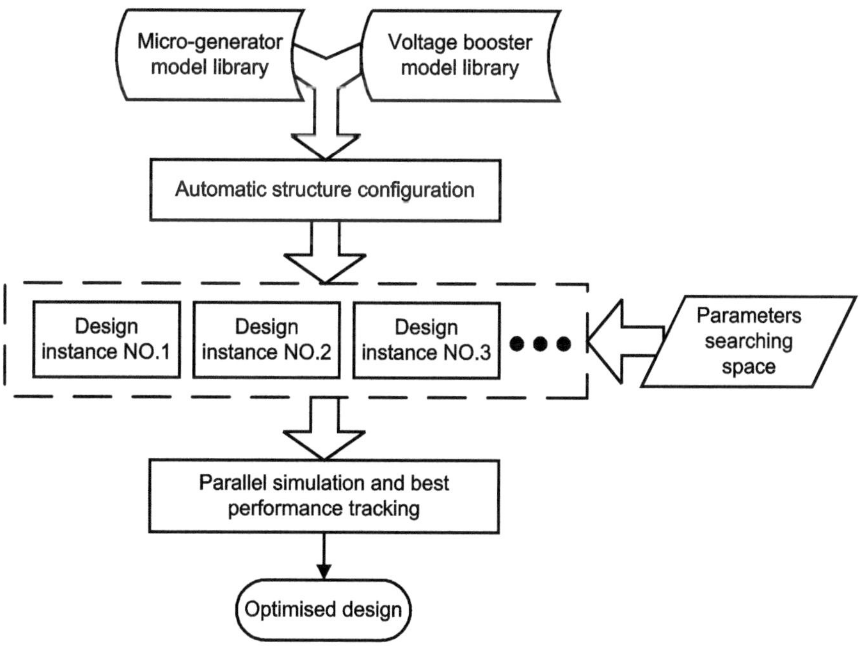

Fig. 2.14 Implementation of the proposed energy harvester design flow in VHDL-AMS

To validate the effectiveness of the proposed approach to improve energy harvesting efficiency, the following simulations and experimental measurements have been carried out.

2.5.1.1 Original Design

This combines type II micro-generator with a five-stage Dickson VM. The used VM has been reported in literature as optimal configuration [18]. However, in the original design these two parts are optimised separately, which is quite common in existing energy harvester design approach. Parameters of original design are listed in Table 2.4.

Table 2.4 Parameters of original energy harvester

Micro-generator	
Wire diameter (μm)	16
Coil thickness (mm)	1.3
Coil radius (mm)	2.45
Voltage booster	
VM configuration	five-stage Dickson
Capacitor values (C1–C5, μF)	47,150,150,47,150

2.5.1.2 Optimised Design

This has been obtained using the proposed design flow (Fig. 2.14). Table 2.5 gives the new micro-generator and voltage booster parameters.

Table 2.5 Parameters of optimised energy harvester

Micro-generator	
Wire diameter (μm)	25
Coil thickness (mm)	1.3
Coil radius (mm)	2.0
Voltage booster	
VM configuration	Three-stage Dickson
Capacitor values (C1–C3, μF)	100,100,47

The impact of these values on improving the energy harvester performance has been validated in both simulation and experimental measurements. According to the optimisation result, a new coil has been ordered from Recoil Ltd, UK [12], and replaced the original one for testing (see Fig. 2.15).

Simulation and experimental waveforms of the original and optimised design are shown in Fig. 2.16. The impact of using the supercapacitor model in Fig. 2.8 instead of an ideal capacitor has also been investigated. As can be seen from the figure, there is good correlation between the simulation and experimental waveforms in both of the energy harvester designs, which validates the effectiveness and accuracy of the

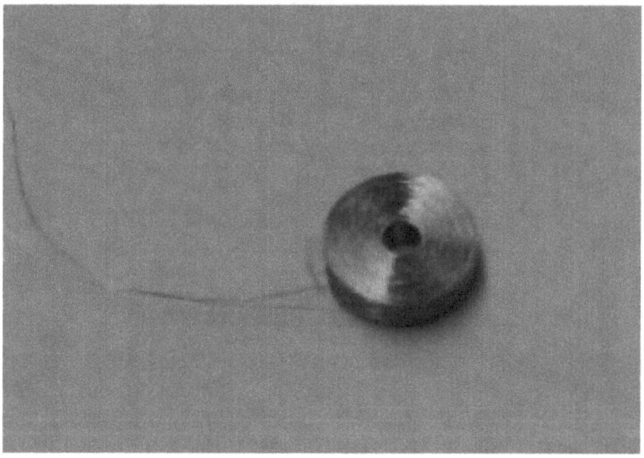

Fig. 2.15 New coil according to optimisation result ($R = 2.0$ mm, $r = 0.5$ mm, $t = 1.3$ mm, $d = 25\ \mu$m)

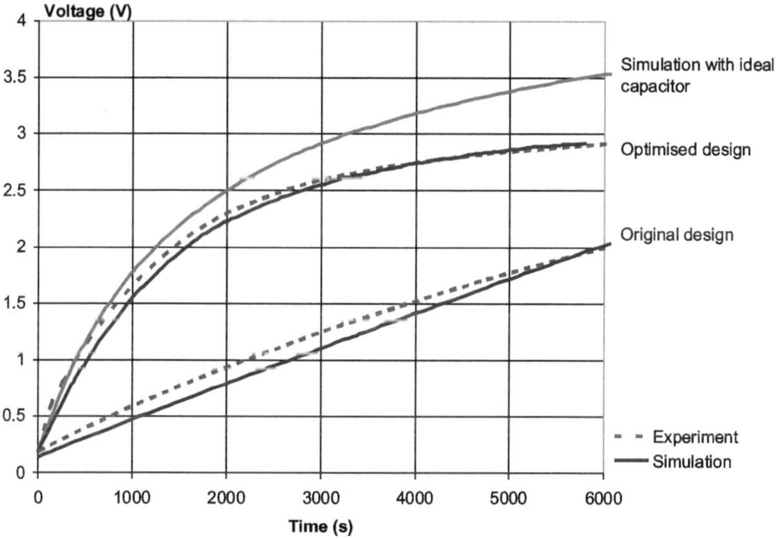

Fig. 2.16 Simulation and experimental waveforms of original and optimised energy harvesters

proposed design flow. The energy harvester from original design can charge the supercapacitor to 2 V in 6000 s while the optimised design only uses 1500 s, which represents a 75% improvement.

2.5.2 Genetic Optimisation

This section demonstrates another possible optimisation method to improve the energy harvester efficiency. Figure 2.17 shows that in the proposed approach, not

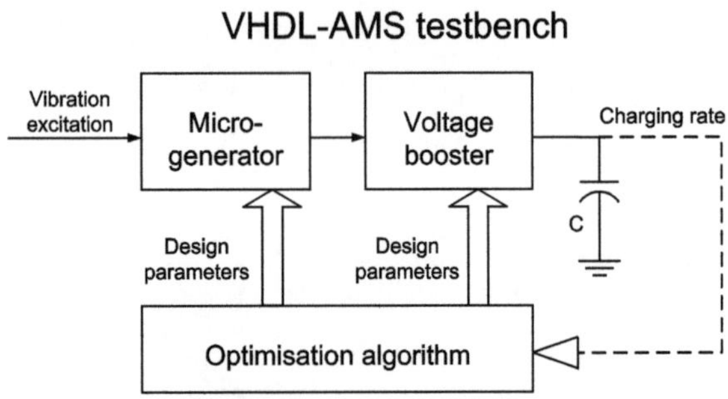

Fig. 2.17 Integrated performance optimisation in VHDL-AMS testbench

only the energy harvester model but also the optimisation algorithm is implemented in a single VHDL-AMS testbench. The parameters used for the optimisation are from both the micro-generator and the voltage booster. The optimisation object is to increase the charging rate of the supercapacitor. The optimisation algorithm generates design parameters to the model and obtains the charging rate through simulation. The optimisation loop runs continuously until the design parameters reach an optimum.

A supercapacitor of 0.22 F has been used in the performance optimisation experiment. The micro-generator parameters that can be optimised are the number of coil turns (N), the internal resistance (R_c) and the outer radius (R). The voltage booster circuit is the voltage transformer described in Section 2.3.2.2 (Fig. 2.7). The parameters are the number of turns and the resistance of primary and secondary windings. For proof of concept, we employed a genetic algorithm (GA) [7] to optimise the energy harvester with a voltage transformer booster. The implemented GA has a population size of 100 chromosomes. Each chromosome has seven parameters (three from the micro-generator and four from the voltage booster). The crossover and mutation rate are 0.8 and 0.02, respectively. Other optimisation algorithms may also be applied based on the proposed integrated model. The "unoptimised" model parameters are given in Table 2.6.

Applying the proposed modelling and performance optimisation, Table 2.7 gives the new micro-generator and voltage booster parameters which are referred to as "optimised" design. The impact of these values on improving the charging of the supercapacitor is shown in Fig. 2.18. As can be seen from the simulation results,

Table 2.6 Parameters of unoptimised energy harvester

Micro-generator		
Outer radius of coil (R)		1.2 mm
Coil turns (N)		2300
Internal resistance (R_c)		1600 Ω
Voltage transformer		
	Resistance(Ω)	No. of turns
Primary winding	400	2000
Secondary winding	1000	5000

Table 2.7 Parameters of GA-optimised energy harvester

Micro-generator		
Outer radius of coil (R)		1.1 mm
Coil turns (N)		2100
Internal resistance (R_c)		1400 Ω
Voltage transformer		
	Resistance (Ω)	No. of turns
Primary winding	340	1900
Secondary winding	690	3800

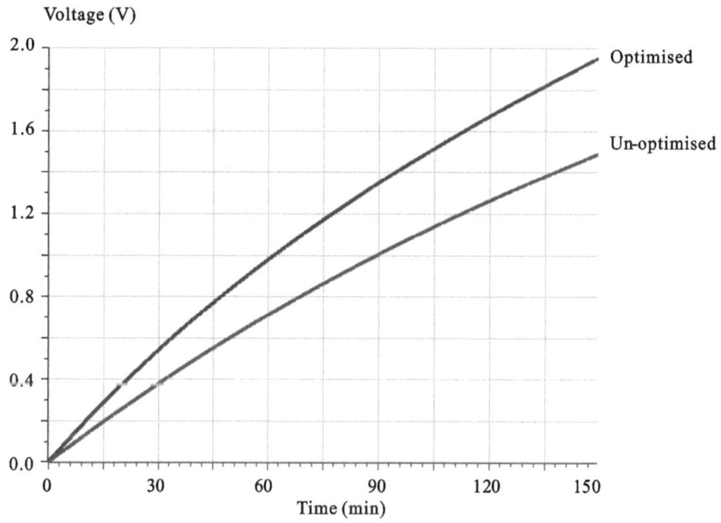

Fig. 2.18 Simulation waveforms of supercapacitor charging by different energy harvester models

in 150 min the unoptimised energy harvester charges the supercapacitor to 1.5 V and the optimised energy harvester reaches 1.95 V, which represents a 30% improvement.

Performance of the developed GA has been further investigated by comparing the power transfer efficiency before and after optimisation. The maximum average power that can be delivered to the electrical domain is about 144 μW, calculated from Eq. (2.10). Table 2.8 lists the average electrical power output from the micro-generator and the voltage transformer. It can be seen that the optimisation improves the efficiency of both the micro-generator and voltage booster, which validates the effectiveness of the developed genetic optimisation.

Table 2.8 Energy harvester power efficiency

	Generated power (μW)	Generator efficiency (%)	Delivered power (μW)	Transformer efficiency (%)	Overall efficiency (%)
Pre-optimisation	26.875	18.66	15.750	58.60	10.94
Post-optimisation	29.250	20.31	19.625	67.09	13.63

2.6 Concluding Remarks

This chapter presents an automated energy harvester design flow that can generate optimised configuration from an existing HDL model library as well as carry out performance optimisation through the employment of a single software platform. The effectiveness of the proposed design flow has been demonstrated by automatic configuration, optimisation and experimental validation of an energy harvester powered by an electromagnetic vibration-based micro-generator. It has been shown that the existing energy harvester design approaches are inadequate because there is a trade-off between different energy harvester components and the optimisation of subsystems in isolation does not lead to a globally optimal design. A new energy harvester has been manufactured according to the outcome from the proposed design flow and experimental measurements of the new device have validated the optimisation results. It has been shown that the outcome from the design flow (configuration and optimisation) can achieve a 75% improvement in the supercapacitor charge rate and the integrated performance optimisation alone may achieve a 30% improvement.

References

1. Amirtharajah R, Wenck J, Collier J, Siebert J, Zhou B (24–28 July 2006) Circuits for energy harvesting sensor signal processing. In: Design Automation Conference, 2006 43rd ACM/IEEE, Anaheim, CA 639–644
2. Beeby S, Torah R, Tudor M, Glynne-Jones P, O'Donnell T, Saha C, Roy S (2007) A micro electromagnetic generator for vibration energy harvesting. J. Micromech. Microeng. 17(7):1257–1265
3. Beeby SP, Tudor MJ, White NM (2006) Energy harvesting vibration sources for microsystems applications. Meas. Sci. Technol. 17(12):R175–R195
4. von Buren T, Mitcheson P, Green T, Yeatman E, Holmes A, Troster G (2006) Optimization of inertial micropower generators for human walking motion. Sens. J. IEEE 6(1):28–38
5. Corporation MG (July 2004) SystemVision User's Manual. Version 3.2, Release 2004.3
6. El-Hami M, Glynne-Jones P, White N, Hill M, Beeby S, James E, Brown A, Ross J (2004) Design and fabrication of a new vibration-based electromechanical power generator. Sens. Actuators A: Phys. 92(1–3):335–342
7. Mitchell M (1996) An Introduction to genetic algorithms. MIT Press, Cambridge, MA
8. Mitcheson P, Green T, Yeatman E, Holmes A (2004) Architectures for vibration-driven micropower generators. J. Microelectromech. Syst. 13(3):429–440
9. Mitcheson P, Miao P, Stark B, Yeatman E, Holmes A, Green T (2004) MEMS electrostatic micropower generator for low frequency operation. Sens. Actuators A: Phys. 115(2–3):523–529
10. Nelms R, Cahela D, Newsom R, Tatarchuk B (14–18 March 1999) A comparison of two equivalent circuits for double-layer capacitors. In: Applied Power Electronics Conference and Exposition, 1999, vol. 2. APEC '99. Fourteenth Annual, Dallas, TX 692–698
11. Ottman G, Hofmann H, Bhatt A, Lesieutre G (2002) Adaptive piezoelectric energy harvesting circuit for wireless remote power supply. Power Electron. IEEE Trans. 17(5):669–676
12. Recoil Ltd, UK (Sept. 2008) http://www.recoilltd.com/index.htm
13. Saha CR, O'Donnell T, Loder H, Beeby S, Tudor J (2006) Optimization of and electromagnetic energy harvesting device. Magnetics, IEEE Trans. 42(10):3509–3511

14. Shao H, Tsui CY, Ki WH (21–24 May 2006) A charge based computation system and control strategy for energy harvesting applications. In: Circuits and Systems, 2006. ISCAS 2006. Proceedings. 2006 IEEE International Symposium, Island of Kos, Greece 2933–2936
15. Siebert J, Collier J, Amirtharajah R (8–10 Aug. 2005) Self-timed circuits for energy harvesting AC power supplies. In: Low Power Electronics and Design, 2005. ISLPED '05. Proceedings of the 2005 International Symposium, San Diego, CA, 315–318
16. Torah R, Beeby SP, Tudor MJ, O'Donnell T, Roy S (2006) Development of a cantilever beam generator employing vibration energy harvesting. In: Proceedings of the 6th Int. Workshop on Micro and Nanotechnology for Power Generation and Energy Conversion Applications, Berkeley, CA 181–184
17. Torah R, Glynne-Jones P, Tudor J, O'Donnell T, Roy S, Beeby S (2008) Self-powered autonomous wireless sensor node using vibration energy harvesting. Meas. Sci. Technol. 19(12):ISSN 1361–6501
18. Torah R, Tudor M, Patel K, Garcia I, Beeby S (28-31 Oct. 2007) Autonomous low power microsystem powered by vibration energy harvesting. In: Sensors, 2007 IEEE, Atlanta, GA, 264–267
19. Tsui CY, Shao H, Ki WH, Su F (24–27 Jan. 2006) Ultra-low voltage power management circuit and computation methodology for energy harvesting applications. In: Design Automation, 2006. Asia and South Pacific Conference, Yokohama, Japan 96–97
20. Wang A, Chandrakasan A (2005) A 180-mv subthreshold fet processor using a minimum energy design methodology. IEEE J. Solid State Circuits 40(1):310–319
21. Williams C, Shearwood C, Harradine M, Mellor P, Birch T, Yates R (2001) Development of an electromagnetic micro-generator. Circuits, Dev. Syst. IEE Proc. 148(6):337–342
22. Yan H, Macias Montero J, Akhnoukh A, de Vreede L, Burghartz J (17–18 Nov. 2005) An integration scheme for RF power harvesting. In: Proceedings STW Annual Workshop on Semiconductor Advances for Future Electronics and Sensors, Veldhoven, The Netherlands 64–66

Chapter 3
Simulation of Ultra-Low Power Sensor Networks

Jan Haase, Joseph Wenninger, Christoph Grimm, and Jiong Ou

Abstract This chapter gives an introduction into the importance of power analysis and power harvesting in wireless sensor networks. It additionally gives an overview over simulation techniques for this topic and an overview of related tools and methodologies, but mainly focusing on SystemC and SystemC AMS and extension libraries.

Keywords Simulation · SystemC · Wireless sensor networks · Ultra-low power · Routing · TLM · OFDM · TUV buildimg block library

3.1 Introduction

Energy harvesting is a very important method to implement long running systems, which should not need to be recharged at power plugs. That is, for instance, because they have to be always at places, where there just is no power plug nearby or because they are mobile. It could also be difficult or even impossible to connect the system with a wire to other parts. Additionally to energy harvesting the need arises to reduce the systems' overall power consumption, because there could be periods of times when the harvesting mechanism does not work, for instance, if the harvester gains energy from vibrations and there are moments, when there is just not enough movement, so this period of time has to be bridged.

At first a look on what comprises an ultra-low power wireless sensor network should be had and what are the properties of the components and the system at large. As an example we are going to use a weather station as shown in Fig. 3.1. As can be seen the overall system consists of various sensor nodes, denoted as "H","L" and "T". Those nodes operate individually to achieve their measurement tasks and cooperate to build up a wireless mesh network for communication purposes. The example uses one dedicated sink node "S", which receives all the sensor data

J. Haase (✉)
Vienna University of Technolgy, Institute of Computer Technology, Gusshausstrasse 27-29, 1040 Vienna, Austria
e-mail: haase@ict.tuwien.ac.at

T.J. Kaźmierski, S. Beeby (eds.), *Energy Harvesting Systems*,
DOI 10.1007/978-1-4419-7566-9_3, © Springer Science+Business Media, LLC 2011

gathered within the network. Important properties of the system, especially for large quantities of sensors, are

- cheap nodes;
- small nodes;
- self-configuration (especially network topology);
- robust nodes and communication;
- safe operation; and
- wireless communication using standards, e.g. ZigBee (IEEE 802.15.4; 868 MHz, 915 MHz, 2.4 GHz).

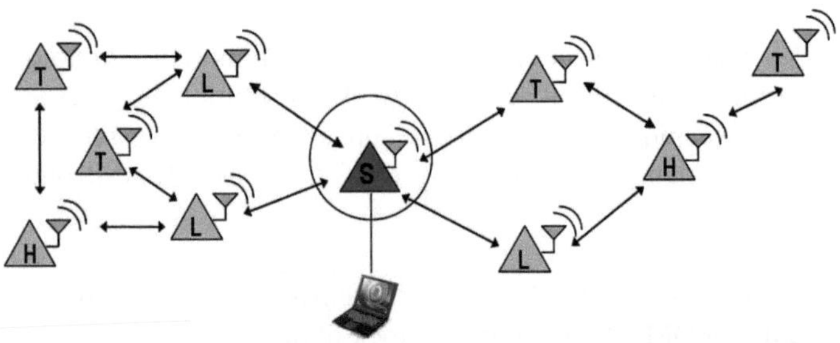

Fig. 3.1 Distributed wireless weather station, "H" are humidity sensors, "L" are light sensors, "T" are temperature sensors, "S" is the sink node

If we go further and want our sensor nodes to be energy autonomous additionally to the properties mentioned above appear the following things have to be considered:

- Usage of energy harvesters
- Usage of capacitors or batteries
- Average power consumption reduced to some microamperes
- Reduction of duty times (most time "off") (duty cycles $< 1\%$)
- Low data rates (1–50 kbps or even below)
- Usage of ultra-low power circuits
- Resource management/power saving: switching off of "unneeded" resources

Now that we have identified the main properties of interest we have to identify the factors and functional properties which have the highest impact on them. Those factors are, for instance, the acceptable latency which can be within some milliseconds to several days or the transmission rate, mostly within the range of some kilobytes per second. For classic wireless networks and for mesh networks the distance between the nodes is also important, since it has an impact on the needed transmission power. The range for this property is typically from some centimeters to various kilometers. Depending on different scenarios of the surroundings and possible changing positions of the nodes the average and maximum power consumption have to be evaluated. The best way to do this is to create a very precise model of a lot of scenarios and to simulate the entire system on system level, not just one node. How to make it easier and how to speed up the creation of such system-level simulations is shown in Section 3.4. It is also shown there how to speed up the simulation

Fig. 3.2 Example sensor node [4]

itself by varying the accuracy and design level of the model for various parts of the system. The needed lifetime without an external power source is a crucial factor, which is heavily correlated to the power consumption. For a given lifetime the two parameters consumption and production (harvesting or storage (battery)) have to be leveled accordingly. Last but not least related to the costs is the quantity of nodes going to be used; it could be in the range from tens to many hundred thousands of nodes. The higher the quantity, the more important the price of the hardware in use and the costs of the development process.

Now that the properties of the overall sensor network are specified, it is important to look at the components of the sensor or communication nodes itself. A node consists of four virtual distinct (some can be within the same physical die or chip) parts (Fig. 3.2):

- Power supply, i.e. battery or harvester (solar cell, induction loop, MEMS, etc.)
- Sensing unit
- Communication unit
- Controller

In a lot of cases the sensing unit does not only have to cope with digital input values, which is more or less easy to handle, but has to measure and evaluate analog signals, for instance, like in the weather station mentioned earlier (Fig. 3.1), there are temperatures, pressures, light intensities, which are inherently analog. Therefore it contains analog filters, analog digital converters (ADC) and digital filter components for handling the signal. As already mentioned parts of this can be moved into other components. For instance, the digital filters could be integrated into the main controlling unit, if it is not a simple microprocessor or microcontroller, but a full-fledged digital signal processor (DSP). The ADC does not have to be a separate part on its own either, it can also be part of the controller, but it is still there of course.

The communication unit is needed for transmitting and receiving data to/from other nodes. It is partly split into software running on the microcontroller and the transceiver, which does the handling of the physical transmission over the air. As we are going to see later in Section 3.2, there are ways to optimize this architecture. For instance, there is the possibility that the media access control (MAC) is implemented within the hardware and/or software layer of the main microcontroller or it can be moved into the transceiver part. This alone opens the possibility of various optimizations, especially related to the power consumption of the whole system.

3.1.1 Scenario Analysis

As already mentioned in Section 3.1, there are many properties and factors which have to be taken into consideration to allow an educated decision-making process for the overall system design. Since the number of possibilities is usually very large to an extent, where it is impossible to enumerate all disjunct property sets, the need arises to define specific scenarios and use cases for the system under planning and delivering. The fundamental result of those scenario analyses is an equation which has to be fulfilled at all times. It is

$$available\ power(t) >= required\ power(t).$$

If this fundamental rule is not fulfilled the system will fail in a given scenario. For estimating the worst cases and the boundaries of the equation, for instance, a Monte Carlo analysis can be done to support the creation of worst-case scenarios, which should still be possible to be handled by the system without failure. The functions on both sides of the equation are depending on various time-dependent properties. Possible dependencies are

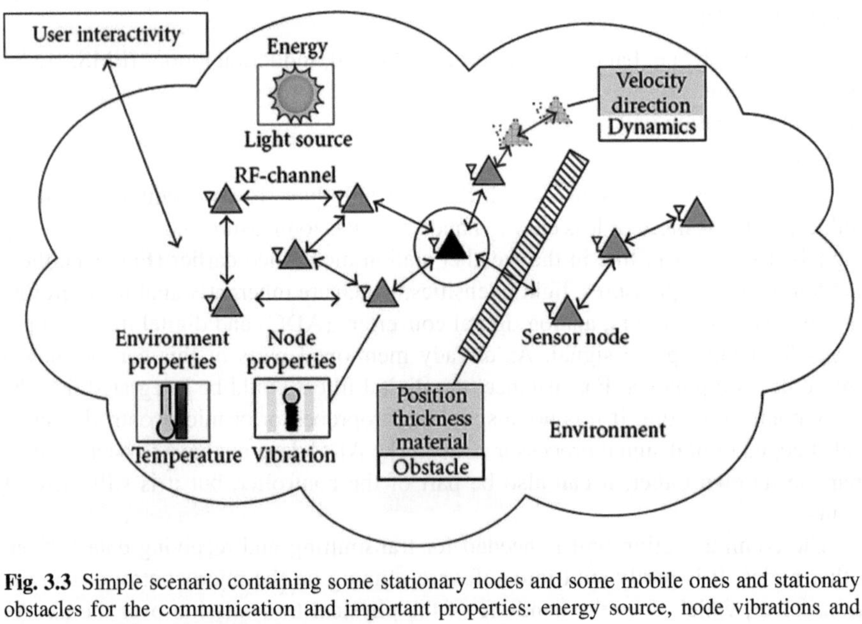

Fig. 3.3 Simple scenario containing some stationary nodes and some mobile ones and stationary obstacles for the communication and important properties: energy source, node vibrations and environmental parameters like the temperature

$$available\ power(t) = f(environment,\ harvester,\ battery)$$

and

$$required\ power(t) = f(architecture,\ environment, \ldots)$$

The more the scenarios are considered and the more they are realistic models of the real world, the better the assessments are. A very simple scenario for an analysis of sensor nodes within a restricted area and given environmental parameters, including possible obstacles and node movements, can be seen in Fig. 3.3. In the end it is important for optimal functionality and reliability that the harvesters match the sensor network consisting of nodes and their environment. Based on this evaluation it has to be considered which optimizations are possible, needed, useful and financially feasible.

3.2 Design Space and Design Issues

In this section we are going to shed some light on design issues and possible design decisions to explore the whole design space. It is not planned to be an exhaustive enumeration, but should give hints on what to consider related to the communication part of a sensor node to establish energy efficiency.

3.2.1 Issues

One point is the architecture of the underlying communication network itself. For instance, it could be decided whether the communication is conducted via a centralized hub node. This star topology, for instance, is low power at low distances. It could be fragile though, since if the hub node fails itself or for some nodes the communication with the hub fails, e.g. because an obstacle moved between the leaf nodes and the center node, important messages could easily get lost. An extension to this topology could be to create a star of stars, where there are local hub nodes handling communication within their vicinity and communicate with other subranges of the network via a higher order hub node. In this topology the communication protocol has to support multiple hops for message deliveries. The advantage is that the system still is low power, although higher distances can be bridged. The disadvantage is the higher complexity of the protocol used. The topology described here uses static routing, so if one of the hub nodes fails, or the link between a lower and a higher order hub node fails, a whole network segment can become disconnected and thus important, possibly, time critical data cannot be transmitted appropriately. It is important to note that the various hub nodes have higher energy consumptions than the leaf nodes, since they are always between two nodes communicating with each other. Therefore the power supplies (either harvesters or batteries) for those nodes have to be dimensioned to provide higher power capacities to keep the whole network alive, while the power supplies for leaf nodes can be dimensioned smaller

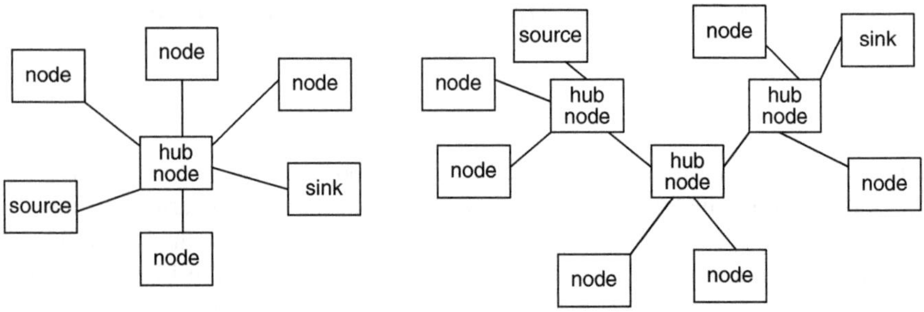

Fig. 3.4 Star topologies: on the *left-hand side* the whole network is just one big star and on the *right-hand side* the image shows a star with some nodes creating a sub-star. The latter topology is similar to a *tree structure*

concerning capacity. Figure 3.4 shows examples for this basic network topologies. This essentially leads the designer to evaluate which architecture (topology), which used protocols for the higher levels and which media access scheme suits a given use case best.

There are ways to increase the reliability of the overall system while reducing the load on every single node, to the point where all communication nodes have an approximately equal load concerning transmissions. In this case intelligent routing, e.g. mesh routing, can be applied. For this system to work, the protocol has to reliably detect the distances between various nodes, the hops needed for a communication between a given source and a given sink and which alternative routes exist. It is important too to keep track of changes of the network topology, e.g. failed nodes, changed distance vectors, new nodes. This knowledge allows various optimizations of the message delivery. For instance, each node (hop) could retransmit a received message to the nearest node (the node which can be reached with the least transmission power) in line to the sink node. On the other hand, the communication can be trimmed to use the least count of hops in between a source and a sink. If there exist roughly equal routes related to costs, they could be used in an alternate pattern to distribute the power consumption equally across nodes. The cost function for a route is roughly $cost(route) = f(node\ energy\ hop\ count,\ distances,\ node\ energy\ of\ next\ hop(s),\ \ldots)$. This kind of routing can increase the reliability of the network and level the power consumption out between nodes. The reliability increases, since if one communication link fails there might be an alternative route between two points. Each advantage comes with a drawback though. The implementation might need more hardware, which increases the power consumption of each single node. It might also increase the traffic within the network, since the topology has to be always detected accurately to make the routing work, so in either fixed distances of time or depending on some parameters in varying intervals the topology has to be reevaluated. Those parameters could be the knowledge of directions and speeds of node movements or knowledge of energy levels. This keeping of the topology information up to date increases power consumption, since communication additional to the application

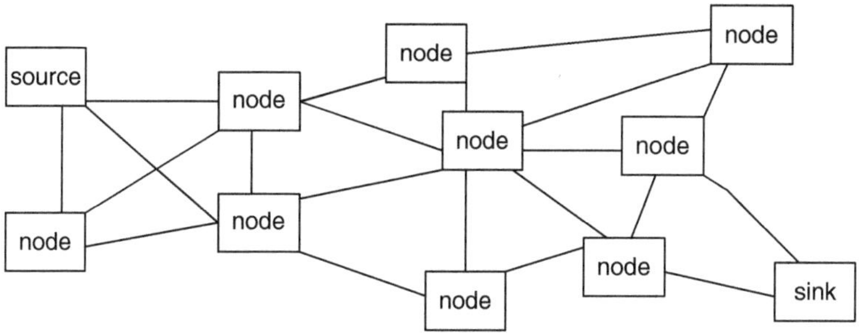

Fig. 3.5 Mesh topology: here more than one path from any given sender to any given sink is possible

payload is required. As can be seen there might be a trade-off between reliability and power consumption, which has to be appropriately considered in the decision-making process. See Fig. 3.5 for an example, how such a more reliable topology may look like.

As already mentioned the updating of the topology needs energy, although this increase can be kept fairly small, depending on the environment and the use cases and the optimizations done. For instance, there is the possibility to provide MAC layer information across the other levels of the networking stack. So the application level can get information about the availability of power or the quality of received signals and so on. If the whole protocol stack is optimized for low power all parts are intertwined concerning power management to ease the decision process of how often to transmit, how to prioritize messages (quality of service, QOS) and which paths to use for communication and how to detect network dropouts early enough to perhaps avoid situations which could violate safety requirements.

Another factor important for the power usage and for the robustness and reliability of the overall system is how the data are encoded on the channel; for wireless sensor networks, this would be the air. There exists a variety of possibilities for modulation schemes, going from very simple to very complex. It can basically be said that the simpler the schema is, the less power it uses, but proportionally (or even exponentially) the robustness of the communication decreases too. The other way also holds true, the more complex a scheme, the more power it needs but the robustness increases with a higher order or even exponentially. Simple schemes are, for instance, ON/OFF-shift keying or more general amplitude shift keying (ASK). Those two can be considered ultra-low power, but very fragile concerning robustness against noise. More complex schemes are frequency shift keying (FSK) or phase shift keying (PSK). In the first case a logical one is represented by a certain frequency on the channel and a logical zero is represented by another one. Other methods could be quadrature amplitude modulation (QAM), where the code points of the signal are part of the complex plain and each code point can represent more than one symbol. FSK and PSK can be considered more robust than ASK but still as low, although not ultra-low, power.

3.2.2 Partitioning

Additional to the environment, the topology, the protocol and the modulation there is another important thing to consider. This is the partitioning between software and hardware.

What does partitioning mean? If we look, for instance, at the receiver part of a transceiver, there are two ways of approaching the implementation:

1. Implementing as much as possible in software
2. Implementing as much as possible in (analog) hardware

The advantage of the first way is that the resulting design is very flexible, the disadvantage though is that the microcontroller is most of the time on, although probably idle, waiting for data to decode. An example for this situation is shown in Fig. 3.6.

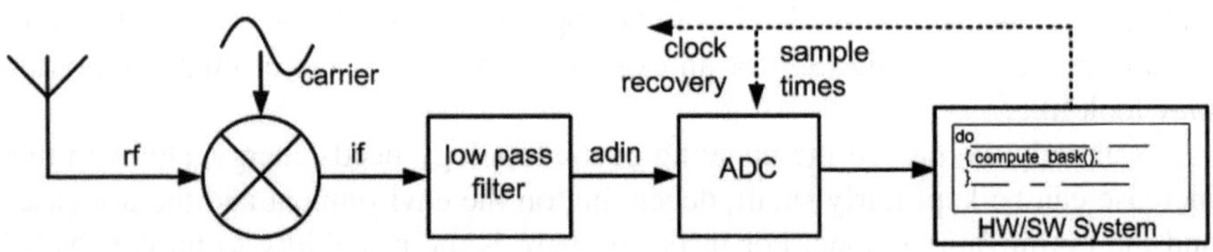

Fig. 3.6 Mostly software-based receiver: the radio frequency (rf) comes through the antenna into the receiver, is mixed with the carrier frequency and afterwards filtered via a low-pass filter. The filter signal is converted to a digital signal and the digital hardware combined with the software does the decoding of the transmitted symbols

The advantage of the mostly analog design is that it needs a lot less power than the other approach. As always there are disadvantages too. The design has a fixed functionality; it is difficult to design, because it has to be first time right and patches cannot be shipped after the roll-out to fix problems in a delivered system. The design is difficult or not at all portable to different technologies. In general the robustness is very low and it is not that easy to reach a high yield during the production part of the system's live cycle.

As always the optimum lies somewhere in between, partitioning does exactly deal with that and relies heavily on tools and methodologies, which allow a fast, more or less exhaustive exploration of the design space, according to given real-world constraints. For instance, the usage of reconfigurable hardware could increase flexibility, while still staying low power, but it increases the efforts during development.

Fig. 3.7 is an illustration of the principle by means of our example. Moving the media access control (MAC) layer into the hardware increases the standby times of the microcontroller, but makes the design less flexible for adaption to other access schemes.

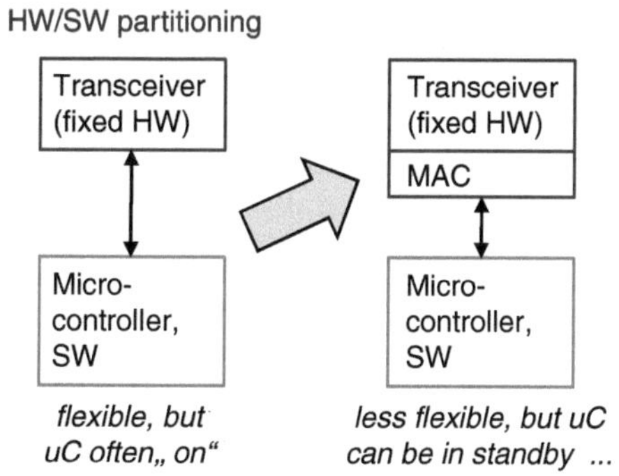

Fig. 3.7 HW/SW partitioning: here the optimization of moving the media access control (MAC) part out of the software part running on a microcontroller (uC) into the transceiver hardware is shown

3.2.3 How to Solve the Issues

As already mentioned it is mandatory to use “ultra-low power” circuits to achieve overall low power systems and switching off parts, currently not needed is the most important of power saving strategies. To get good information to base decisions on, the overall sensor network has to be simulated and this simulation has to show which resources are used at which times and how much energy they use at every single point in time. It is useful if not crucial to employ mixed-level simulations, which encompass architecture functional and network level and which can show the benefits or failed attempts of various cross-level optimizations.

For each level of simulation there are specialized tools available, which eventually if used together can speed up the design process and simulation performance, by varying the detail levels of certain parts of the system. Table 3.1 shows a list of common tools. In the further parts of this section we are going to put our focus on SystemC (Section 3.2.4) and SystemC AMS (Section 3.2.5) showing how those tools can be employed to our specific problem (power simulation) at hand. In Section 3.4 we give an introduction to a communication library for SystemC AMS which has been developed at the Vienna University of Technology to ease the design of high-level simulations of transceiver systems including the transmission channel.

3.2.4 SystemC

SystemC [6] is a class library for C + +, which allows a high-level simulation of digital circuits. It allows simulation of concurrent modules, various digital signal types and of course time. The core of the library is a discrete event (DE) simulation kernel. A good thing about SystemC is that using the same technology software and hardware can be simulated at the same time. There exists a subset of SystemC, which is even synthesizable to, for instance, FPGA (field-programmable gate

Table 3.1 Tool domains

Tool	Appropriate for
SPICE	Circuit level design (RF parts)
VHDL-AMS, Verilog-AMS	Design of AMS subsystems (PLL, ADC, Harvester)
Matlab/Simulink	Functional behaviour
SystemC	Functional behaviour + architecture of HW/SW systems
SystemC AMS extensions (OSCI draft b1)	Functional behaviour + architecture of AMS systems
OMNET++, other extensions	Network function and architecture

arrays). In SystemC classes are comparable with hardware components and there are few restrictions on what a module can do. It is possible to create classes representing basic adders up to implementations of a complete arithmetic and logic unit (ALU) and so on. Modules can be created hierarchically, so, for instance, within the mentioned ALU various adders and other components can be instantiated to implement the ALU functionality. A module (component) can contain any C ++ code, so software parts are just an implementation detail of one module. This allows easy partitioning, since, for instance, a fast Fourier transformation (FFT) module can be realized as hardware or as a software routine running on a small embedded microcontroller later on. Fig. 3.8 visualizes the concept of component reuse and hierarchical modules.

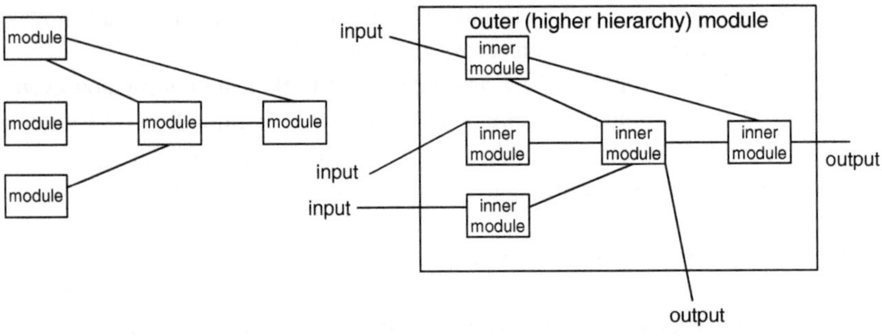

Fig. 3.8 SystemC modules

SystemC has even more advantages. With its signals it is possible to model physical signals (wires) or abstract more lines together into a bus. Signals in SystemC can be either logic values or more complex data types (integers, doubles, etc.). An extension to bundling signals to buses is applied in the transaction-level modeling (TLM) [2] approach. In this approach, which is implemented in the OSCI TLM 2.0 [5] library the communication is split up into various phases and handles larger parts of the communication as some kind of black box. The communication is simulated within a model of a memory mapped bus, which is often used with ASICs to link various components together, e.g. the central processing unit (CPU) core with peripherals. It is possible to define own phases to add additional tags to the payload

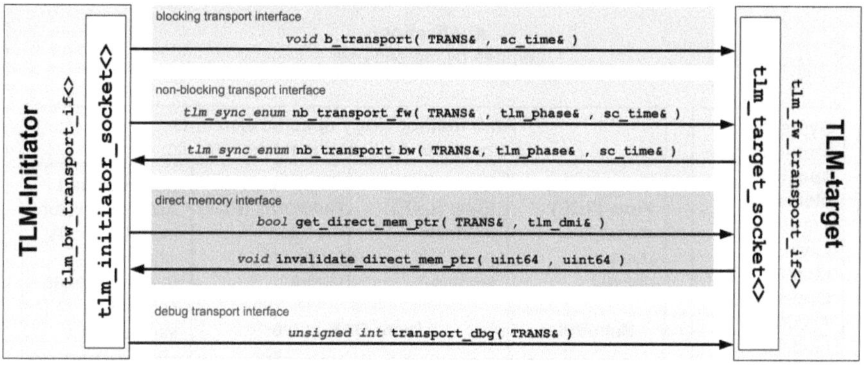

Fig. 3.9 TLM communication: `b_transport` denotes a blocking transport action, `nb_transport` stands for non-blocking transport. `fw` and `bw` stand for *forward* and *backward* communication channels and `if` means that the part is an interface

of the transaction, so this library is very flexible and extensible and should be able to fit almost all purposes while still allowing fast(er) simulation. Essential phases directly supported by the library are request initiation, addressing, reading/writing date and finishing up the transaction. With TLM blocking and non-blocking communication can be simulated. Fig. 3.9 shows how a typical communication on the bus is performed. In TLM the communication is not modeled as a communication between equal peers. There are two types of communication partners. They are initiators and targets. The former one can initiate requests to any target connected to it and handles the communication. The target itself cannot start a transaction, it has to wait for an initiator to request a communication. A typical example for an initiator is the model of a processor, while a typical target would be a random access memory (RAM) module. A module can implement both types though. For instance, an universal receiver and transmitter (UART) can implement with a target socket on the side to the processor and can have a target as well as an initiator socket on the "serial" protocol side.

SystemC together with TLM is capable of creating high-level functional simulations as well as "loosely timed" as down to cycle-accurate simulations of hardware (and software) components.

3.2.5 SystemC AMS

SystemC AMS is an extension library to SystemC, which adds additional models of computation (MoCs) to the simulation kernel. As can be seen in Fig. 3.10 the analog mixed signals (AMS) extension is located above a synchronization layer, and supports different ways of modeling analog parts, which communicate via the synchronization layer with the DE kernel, the digital part so to say. The electrical linear networks (ELN) MoC allows the creation of analog parts, consisting of resistors, capacitors, coils, fixed voltage/current sources or remote controlled sources.

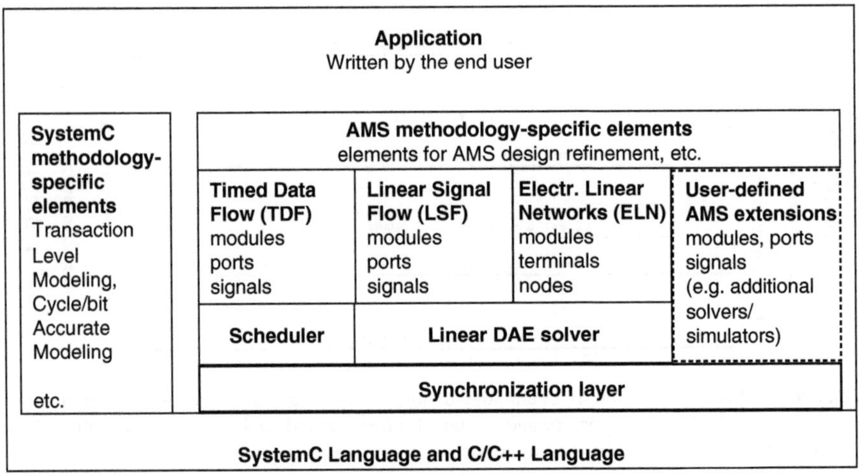

Fig. 3.10 Layer structure of SystemC and SystemC AMS [7]

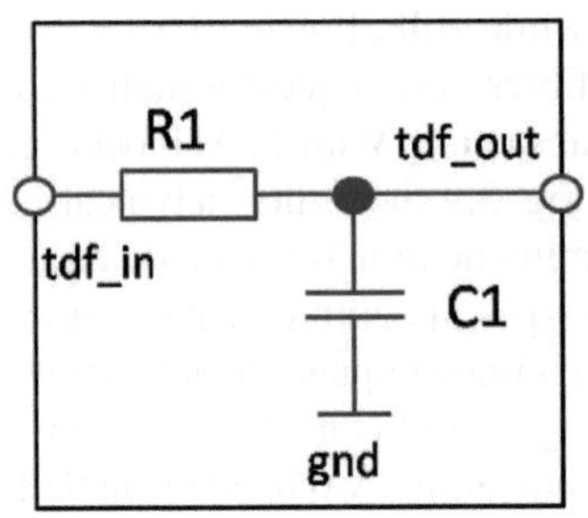

Fig. 3.11 Electrical low-pass filter

Internally it creates a linear differential equation system, which is then solved during simulation, with the given inputs to the network as parameters for the equations. In Fig. 3.11 an example for such a network shows a basic first-order low-pass filter, consisting of just a resistor and a capacitor. Listing 3.1 shows the implementation of this module. The module has timed data flow (see below) input and output signals and models the functionality in a way that the structure of the network is completely obvious, although the network is described in C++. For more complex things this kind of input could be tedious; therefore, other MoCs, which will soon be described, exist to scratch the itch.

The linear signal flow (LSF) is a higher abstraction. In this case the equation system is built from logical blocks, for instance, adders, integrators or differentials.

Last but not least there is the timed data flow (TDF). With this MoC modules can be created containing any C++ code to describe the functionality. The module's processing function is called at given or evaluated time intervals. Each module can be assigned an input and/or output data rate, which has an effect on the scheduling of the whole TDF system, called cluster. The fundamental cluster period, which cannot

Listing 3.1 Software implementation of the low-pass filter as an ELN module

```
SC_MODULE(lp_filter_eln)
{
  sca_tdf::sca_in<double>  in;
  sca_tdf::sca_out<double> out;
  sca_eln::sca_node in_node, out_node; // nodes
  sca_eln::sca_node_ref gnd;           // reference
  sca_eln::sca_r      *r1;             // resistor
  sca_eln::sca_c      *c1;             // capacitor
  sca_eln::sca_tdf2v  *v_in;           // converter TDF->U
  sca_eln::sca_v2tdf  *v_out;          // converter U->TDF
  SC_CTOR(lp_filter_eln) {
    v_in = new sca_eln::sca_tdf2v(âĂIJv_inâĂİ, 1.0);    //
        scale factor 1.0
      v_in->ctrl(in); v_in->p(in_node); v_in->n(gnd);
    r1 = new sca_eln::sca_r(âĂIJr1âĂİ, 10e3);           // 10
        kOhm resistor
      r1->p(in); r1->n(out_node);
    c1 = new sca_eln::sca_c(âĂIJc1âĂİ, 100e-6);         // 100
        uF capacitor
      c1->p(out_node); c1->n(gnd);
    v_out = new sca_eln::sca_v2tdf(âĂIJv_outâĂİ, 1.0); //
        scale factor 1.0
      v_outl->p(out_node); v_outl->n(gnd); v_out->ctrl(out);
  }
};
```

be changed during the simulation, is detected and set up correctly during the elaboration phase, which is a step between setting up the whole system and connecting all signals and the actual start of the simulation. Fig. 3.12 shows a simple cluster consisting of four functional nodes and a delay element. As can be seen, the module `ipl` generates two tokens at each given invocation and the node `dec` consumes always two data values at each invocation. In between is the node calculating the function `f1`, which consumes and produces one value at each invocation; therefore, within one cluster period the processing function of f1 is invoked twice. `dec` itself just produces one token and f2 consumes one token, so `f2` is only evaluated once per cluster period. Since the given example contains a loop there has to be a dead time or delay element to break the direct loop, otherwise the simulation could not progress, since the cycle would be evaluated endlessly during one simulation step. If there would not be a loop, there would not be the need for a delay element per se, although it might arise the need, depending on which physical behaviour is going to be modelled and how accurately it is modelled, e.g. settling times could be simulated with delays.

An example TDF module can be seen in Table 3.2. In this example we highlight what is the important stuff and split the module up according to the functionality of the source line to give a better overview of the semantics.

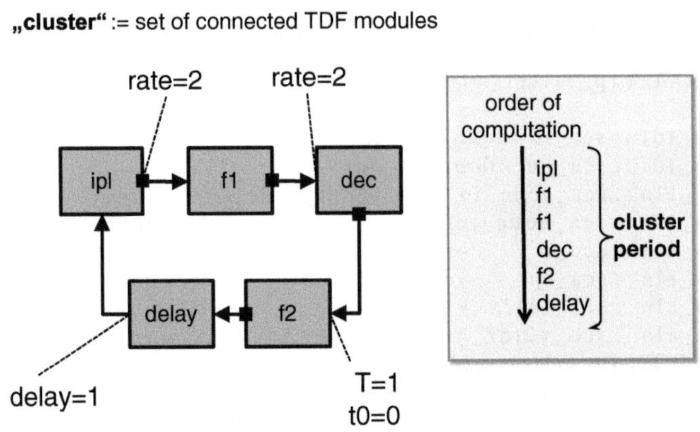

Fig. 3.12 Example cluster (*left*) with its execution schedule (*right*) due to the different number of tokens produced, respectively consumed

Table 3.2 Implementation of a parallel to serial converter with TDF. The different section of the module are separated and described

Module function	Sample code

TDF module: primitive module	`SCA_TDF_MODULE(par2ser)`
	 `{`
	 `sca_tdf::sca_in<sc_bv<8> > in;`
	 `sca_tdf::sca_out<bool> out;`

Attributes specify timed semantics	`void set_attributes()`
	 `{`
	 `out.set_rate(8);`
	 `out.set_delay(1);`
	 `out.set_timestep(1, SC_MS);`
	 `}`

Processing() describes computation	`void processing()`
	 `{`
	 `for (int i=7; i >= 0 ; i- )`
	 `out.write(in.get_bit(i), i);`
	 `}`

Standard constructor	`SCA_CTOR(par2ser);` `}`

3.2.6 Network Level Simulation

In Section 3.2.4 and Section 3.2.5 it has been described how to describe and simulate parts, for instance, the nodes of the system in greater details. However, there is the need to simulate the overall network, not only the subsystems, to get a better feeling of the overall power consumption of the complete system, which a node is only a part of. Although work is done at the Vienna University of Technology to simulate the whole network level also with SystemC and SystemC AMS this is not yet ready for prime time, but there are state-of-the-art tools which could be used for the network level and it is even possible to let those tools cooperate with SystemC to simulate nodes at a fine-grained level and the network in an overview-like level. One of these tools is the OMNET++ network simulator [8]. It is a C-based tool based on a discrete event (DE) kernel. It is free for academy and educational use, but a license has to be payed for commercial usage. It contains a graphical network editor and describes networks with an own network topology description language, the NED. It is a very well-suited tool for simulation of sensor networks and allows co-simulation with SystemC. A screenshot of this tool can be seen in Fig. 3.13.

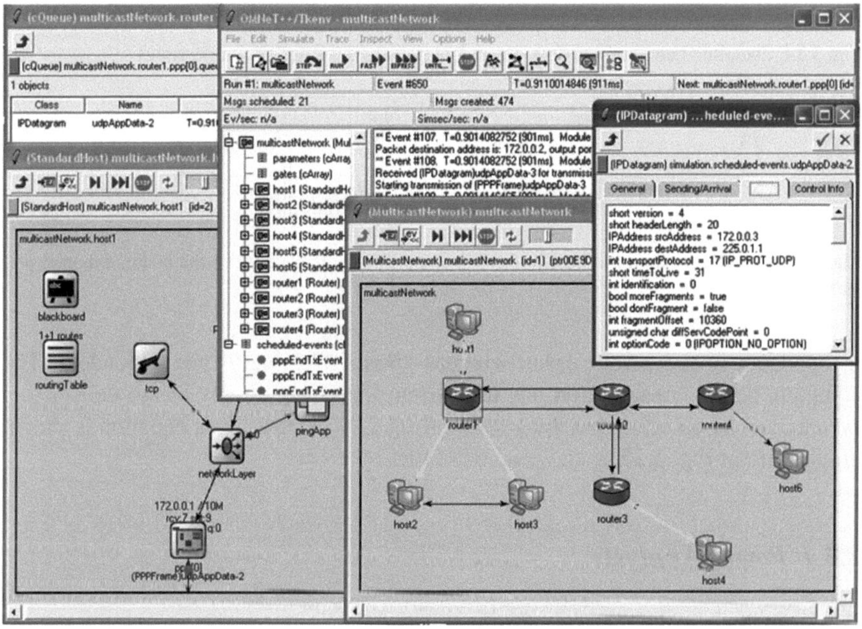

Fig. 3.13 OMNeT++ GUI

3.3 Modeling Strategies for Power Simulation

In Section 3.1 and Section 3.2 the basic problems, goals and possible tools have been described. Now it is time to go further into details on how to do the modelling.

It would be ideal for all models if the power consumption would follow a linear function depending on the components or functions being used. If an ADC is turned on the power consumption always increases the same way and if it is turned off, it decreases always in the same way. The real world is not ideal though, so there is an additional component to the power consumption, which is nasty for modelling purposes. This consumption is dependent on dynamic behaviour and on the history of state changes in the previous (one or more) cycles. So each state/function/component is not characterized by a factor, but by a time-dependent function. Fig. 3.14 shows a visualization. In this figure "State A" could be described as time

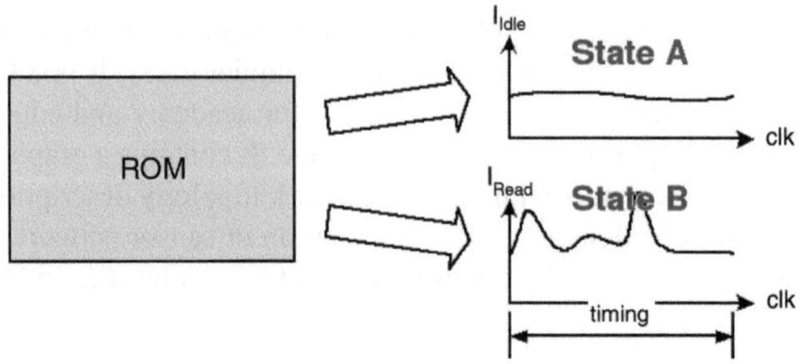

Fig. 3.14 Dynamic behaviour = time-dependent power consumption within states

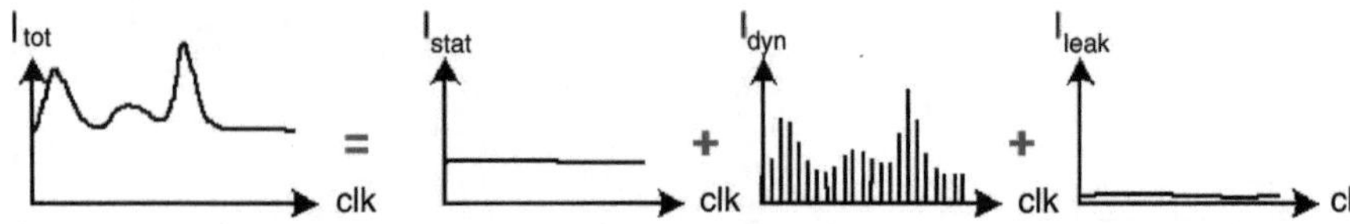

Fig. 3.15 Power equation: the total power consumption I_{tot} is the sum of the static, dynamic and leakage power

independent in the zeroth order, whereas "State B" is very type dependent. The complete power consumption, i.e. the current the power supply has to deliver to a system, can be calculated as $I_{total} = I_{stat} + I_{dynamic} + I_{leak}$. This equation is shown graphically in Fig. 3.15.

3.3.1 Power Analysis

Power analysis is the method of estimating or measuring power usage of a system and drawing conclusion out of the data. The method of doing a power analysis can be described as a cycle, which has some iterations, until the optimization goal has been reached or the decision has to be made, that it is not possible with the given technologies and constraints. As it can be seen in Fig. 3.16 the designer starts with creating a model for the whole system, considering all external constraints. After that the simulation is run in respect of important corner cases, which have to be defined beforehand. During the simulation all usage of each functional block is monitored and somehow logged so this statistical data can be interpreted later on. As a next step the data are used to calculate the overall power consumption, based on multiplication factors or at least linear functions of power consumption per component/function. If the power consumption is not within the allowed range the various parameters (technology, architecture, etc.) have to modified and the whole simulation has to be rerun. It is usually the case that more than one iteration is needed to achieve good results.

Since the analysis needs a lot of runs and the simulation has to cover large time spans to give good results, the performance of the simulation model is essential to keep time frames and schedules. To keep the simulation feasible even for

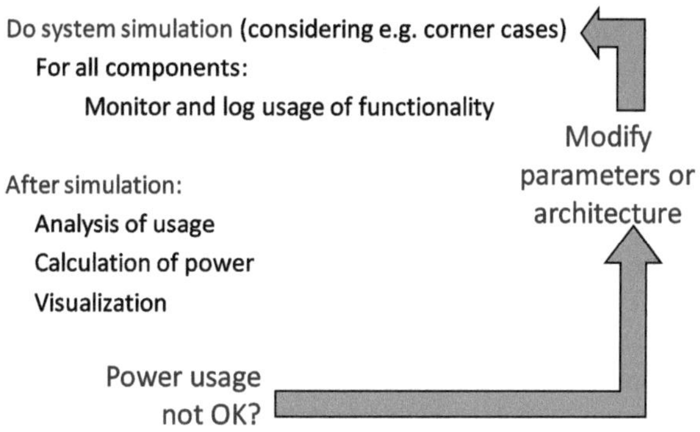

Fig. 3.16 Cycle of analysis

very complex heterogeneous systems consisting of hardware, software and analog mixed signal parts while doing very long time observations the need arises to do concessions:

1. Create models as abstract as possible (but not more abstract than that, otherwise the results are meaningless)
2. Document and communicate which properties and constraints are modelled, which are not and why
3. Be very careful about which information you are tracing and how you are tracing them. Without care traces may become as large as many terabytes, which is not handy for further interpretation tasks.

It is good practice to separate nodes into functional blocks, which can be used independently of each other. This increases the possibility of reuse and has the advantage that various parts can be simulated with different levels of detail. A possible structural dissection would be

- ADC
- Sensor interface
- Transmitter
- Receiver
- Software function calls

For the actual analysis it is a good idea to add a dedicated power meter interface to the simulation (environment), which receives messages from all functional blocks about their current power consumption. To optimize the trace sizes consider trace changes of the consumption and not the consumption itself. This leads to fewer data points and increases the speed of the overall simulation and analysis. The final reports can be of different form, e.g. comma separated value (CSV) files or images showing curves.

3.3.2 Example: State Machine

If a state machine is simulated, in most cases only the functionality is modelled and simulated. For instance, if it has n discrete states s_i, within each state the function f_i is implemented as C or C++ code which takes the input signals and the simulation time as parameters and then computes the state changes and the output vector. In this example of the power simulation a power reporter p_i has to be attached to each state s_i. This reporter is program code which calculates or estimates the power consumption of the functional task and reports it to the overall power meter in the form of $P = U^2/R + \text{const}$. In this case U is the supply voltage, R is an equivalent resistor and const is just a model-specific constant value. The report can be more complex though, for instance, for any kind of (non)linear function. The state machine described here is called state machine with function and power reporting (SMFP).

Creating a power estimation of software functions can also be done with this SMFP model. This is done by creating a timed functional model of the software on the CPU (central processing unit). The estimated runtime (t_i) of each function or procedure f_i is retrieved by measurement or by counting the assembler instructions and adding up the times needed for each instruction. As a next step the software's use of the microcontroller is mapped to the SMFP model. The power of each function can either be measured or be calculated from data sheets, usually the power consumption p_i is nearly constant within each state. The SMFP model reports power usage p_i for execution times t_i of function f_i. In all other cases the SMFP model reports a power consumption p_{sleep} to the power metre monitor.

Even complex functional hardware blocks can be mapped to the SMFP model. For instance, analog digital converters (ADCs), phase-locked loops (PLLs), transmitters or receivers are such modules, which could be bisected and modelled in greater detail or can be modelled as just a functional block with an SMFP. The SMFP template is characterized through circuit simulation, measurements and data sheets to adapt it to the given component.

3.3.3 Modelling the Channel (Air)

In a communication system the transmission channel is of crucial importance. In wireless communication environments this channel usually has the abstract name "the Air". The channel is characterized by reflections, refractions, scattering, fading and in general attenuation of the transmitted signal. In addition the media access scheme is also important; common schemes are TDMA (time division multiple access), FDMA (frequency division multiple access) and CDMA (code division multiple access). From these parameters a matrix can be built, which contains the elements $A_{i,j} = P_{R_{x,i}} / P_{T_{x,j}}$, which describe the attenuation between any given transmitter $T_{x,i}$ and receiver $R_{x,i}$.

3.4 TUV Building Block Library

The Vienna University of Technology (TUV) building block library is a set of SystemC AMS classes for high-level simulation of analog mixed signal modules. It has been published publicly for the first time in conjunction with the ANDRES framework [3], which has been funded by the European Union. It consists of signal sources, signal processing and signal analysis modules. All those modules are based on SystemC AMS. It is published free for private and academic use and can be freely downloaded from [1].

In this section, only a selection of modules and covered areas is shown, as the complete list would fill pages and pages:

- Signal sources
  - Gaussian distributed random numbers
  - Uniformly distributed random numbers
  - Sine
  - Saw tooth wave
  - etc.
- Signal processing
  - Mathematical functions
  - Nonlinearities
  - Basic RF blocks
  - (De)Modulators
  - etc.
- Signal analysis
  - Eye diagram
  - Scatter plot
  - Network analyzer

3.4.1 Introduction to Example Implementation of a Transceiver System

In the following sections (Section 3.4.3, Section 3.4.4, Section 3.4.5, Section 3.4.6) an implementation of an orthogonal frequency division multiplexing (OFDM) transceiver is shown as an example to demonstrate the usefulness of the building block library. It is first implemented as "old style" and then using the library. The system to be built and simulated is depicted in Fig. 3.17. It consists of a generator of test patterns, which are going to be transmitted over the channel with the OFDM

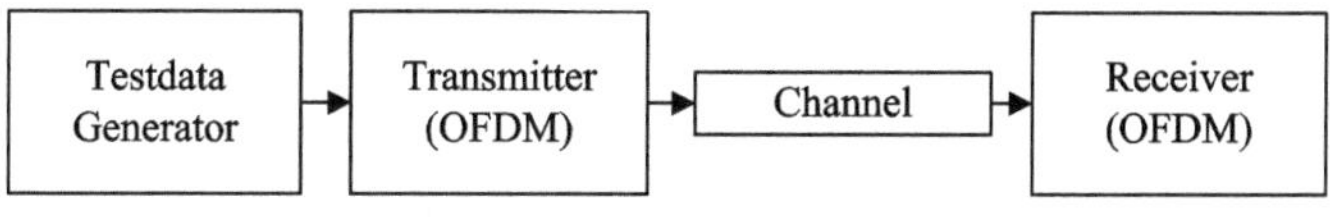

Fig. 3.17 Example transceiver block schematic

modulation. These patterns are handed over to the transmitter module, which does the whole of coding and modulating and handing it over to the channel, which in our case will be a simulation of a transmission through the air with attenuation and noise. This modified signal is afterwards handed over to the transceiver which does the demodulation and returns the bitstream of the received test patterns.

3.4.2 Why Simulate Analog Components

In the context of sensor networks and particularly in low-power sensor networks, transceiver systems are important. The classic sensor networks used wires to connect the various components. This had and still has the advantage that there is a clear line of communication, the wire. Clear bus semantics can be applied to the system, concerning the bus arbitration and electro-magnetic compatibility (EMC). Another advantage of wired buses is that it is easy to estimate the energy consumption of the sending and receiving parts quite easily, because it depends mostly on the system geometry and the resistance of the wires. The big drawback is that the wires have to be placed into building or vehicle structures or for overland networks into the earth or on pylons. With wires on pylons there is the increased danger of lightning strokes, which have to be reduced again through additional lightning protection measures.

Contemporary sensor networks often turn away from wired connections to wireless communication for various reasons. Only short antennas are needed instead of pylons for overland networks and even within objects (buildings, vehicles) there is a big advantage, as no cable connections have to be created between the various nodes of the network. This allows the placement of sensors at points important for data acquisition, even if it would be very expensive or not possible at all to create a reliable wired connection because of the objects geometry or because of moving parts. So it can be said that wireless communication enhances the flexibility of the network. The disadvantage of a wireless sensor network lies in the fact that the power consumption cannot be that easily calculated, because the transmission and receiver unit have no direct point-to-point connections via a wire and therefore the energy is not merely independent of other factors than the geometry of the system. Very important is the influence from other transmission systems sharing the same frequency band (collisions) or having harmonics within the band of interest. Other factors are shielding through stationary, pseudo-stationary moving objects and reflections of signals. Those factors very often lead to retransmissions, more than in wired systems, or to adaption of the transmission power or receiver gain. The energy consumption is most of the time not a linear function. So the best thing to estimate the power consumption and to simulate the system's functional behaviour is to also simulate the analog parts. A wireless sensor network can have either all nodes equal or some nodes with energy constraints and a base station, where energy does not have that big an influence. For instance, the base station is powered via the ordinary power grid, whereas the sensor nodes (agent nodes) are powered by battery or solar cells or other sources of energy, e.g. kinetic energy. In this case for a first estimate you only need to model the agent node precisely, whereas the base station can be modelled roughly, just to simulate the functionality, without the energy simulation part. As we will see in Section 3.4.5 the library is especially useful for this part of the simulation, because it reduces the coding work and provides a consistent handling and good interconnectability of the modules.

There is ongoing work done with the building block library to integrate power annotations and estimations to the modules to enable the user to even evaluate power consumption on a high level, to ease the design space exploration. Design space exploration is a very important part in sensor networks, since it helps finding the optimal solutions between analog hardware, digital hardware and software components and their parameters.

3.4.3 What Is OFDM?

OFDM stands for orthogonal frequency division multiplexing. It is a transmission technique often used in popular modern wireless communication systems; this is why it is used for the example here. The transmitter for OFDM, which is shown in Fig. 3.18, takes a serial stream of binary digits. By inverse multiplexing, these are first demultiplexed into N parallel streams, and each one mapped to a (possibly complex) symbol stream using QAM modulation. An inverse FFT is computed on each set of symbols, giving a set of complex time-domain samples. These samples are then quadrature mixed to passband in the standard way.

Fig. 3.19 shows which parts compose an OFDM receiver. The receiver picks up the signal from an antenna, which is then quadrature mixed down to baseband using cosine and sine waves at the carrier frequency. This also creates signals centred on two times the carrier frequency $2 * fc$, so low-pass filters are used to reject these. The baseband signals are then sampled and a forward FFT is used to convert back to the frequency domain. This returns N parallel streams; each of which is converted to a binary stream using an appropriate symbol detector. These streams are then

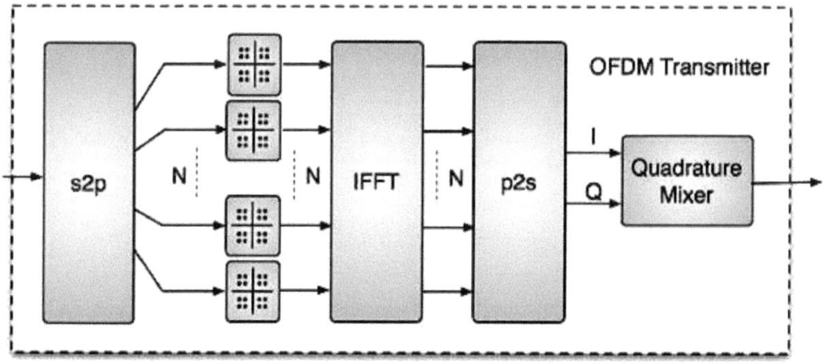

Fig. 3.18 Block schematic of the example OFDM transmitter

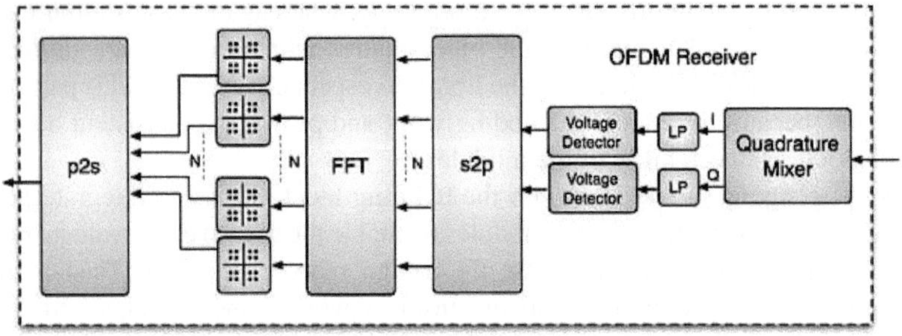

Fig. 3.19 Block schematic of the example OFDM receiver

re-combined into a serial stream, which is an estimate of the original binary stream at the transmitter.

3.4.4 Full-Fledged OFDM Transceiver System

In this section we are going to implement the OFDM transceiver completely without the usage of the building block library or better said, we will show the implementation of the building block modules needed for creating the OFDM transceiver. This chapter focuses on the amount of code needed, not on explaining the complete inner workings; the motto is going to be, let the code speak for itself. An impression of the work should be given, but to not explode the book, just the top-level modules and one four sub-modules of the lower level modules are shown. This is just to get a feeling how much code the implementation of an OFDM transceiver system is. Later on in Section 3.4.5 we are going to show how easy it is to implement the whole system by just using the library instead of reinventing the wheel.

Listing 3.2 OFDM transceiver(1)

```
/*********OFDM_Transmitter**********/
template <int N>
SC_MODULE(ofdm_se){

 public:
  //Ports:
  sca_tdf::sca_in<bool> in;                  // input port
  sca_tdf::sca_out<double> out;              // output port


  //signals
  sca_tdf::sca_signal<bool> sig_pa[N];               //signals on s2p output ports
  sca_tdf::sca_signal<double> sig_real[N];           //signals on q_mapper i output ports
  sca_tdf::sca_signal<double> sig_imag[N];           //signals on q_mapper q output ports
  sca_tdf::sca_signal<double> sig_out_real[N];       //signals on fft real output ports
  sca_tdf::sca_signal<double> sig_out_imag[N];       //signals on fft imag output ports
  sca_tdf::sca_signal<double> sig_out_i;             //signals on p2s i output ports
  sca_tdf::sca_signal<double> sig_out_q;             //signals on p2s q output ports

 private:
  //module instantiation
  s2p<bool,N>* s2p_sub;
  qam_map* qam_mapper_sub[N];
  fft_ifft<N>* ifft_sub;
  p2s<double,N>* p2s_r_sub;
  p2s<double,N>* p2s_i_sub;
  q_mixer_tr* q_mixer_tr_sub;
```

Listing 3.3 OFDM transceiver(2)

```
public:
  //Constructor
  ofdm_se(sc_core::sc_module_name n,double mixer_fc,int
      qam_p_num,double bit_f,int dout_rate,double _amp=1,int
      s2p_or=1,bool mixer_config=false,int p2s_rate=1)
  {

    int mixer_rate=(int)floor(dout_rate*log2(qam_p_num)*
        mixer_fc/bit_f);
    s2p_sub = new s2p<bool,N>("s2p_sub",1);
    s2p_sub->in(in);
    for(int i=0;i<N;i++)
      s2p_sub->out[i](sig_pa[i]);

    for(int i=0;i<N;i++)
      {
        sc_core::sc_module_name nm(((string)"qam_mapper_sub_"
            +(char)(i+65)).c_str());
        qam_mapper_sub[i] = new qam_map(nm ,qam_p_num);
        qam_mapper_sub[i]->in(sig_pa[i]);
        qam_mapper_sub[i]->out_i(sig_real[i]);
        qam_mapper_sub[i]->out_q(sig_imag[i]);
      }

    ifft_sub = new fft_ifft<N>("ifft_sub","IFFT");
    for(int i=0;i<N;i++)
      {
        ifft_sub->in_real[i](sig_real[i]);
        ifft_sub->in_imag[i](sig_imag[i]);
        ifft_sub->out_real[i](sig_out_real[i]);
        ifft_sub->out_imag[i](sig_out_imag[i]);
      }

    p2s_r_sub = new p2s<double,N>("p2s_r_sub",p2s_rate);
    for(int i=0;i<N;i++)
      {
        p2s_r_sub->in[i](sig_out_real[i]);
      }
    p2s_r_sub->out(sig_out_i);

    p2s_i_sub = new p2s<double,N>("p2s_i_sub",p2s_rate);
    for(int i=0;i<N;i++)
      {
        p2s_i_sub->in[i](sig_out_imag[i]);
      }
    p2s_i_sub->out(sig_out_q);

    q_mixer_tr_sub = new q_mixer_tr("q_mixer_tr_sub",mixer_fc
        ,_amp,mixer_rate,mixer_config,0.,0.);
    q_mixer_tr_sub->i_in(sig_out_i);
    q_mixer_tr_sub->q_in(sig_out_q);
    q_mixer_tr_sub->out(out);

  }
};
```

Listing 3.4 OFDM receiver(1)

```
/********OFDM_Receiver**********/

template <int N>
SC_MODULE(ofdm_re){

 public:
  //Ports
  sca_tdf::sca_in<double> in;          // input port
  sca_tdf::sca_out<bool> out;           // output port


  //Signals
  sca_tdf::sca_signal<double> sig_in_i;
  sca_tdf::sca_signal<double> sig_in_q;
  sca_tdf::sca_signal<double> sig_i_lp;
  sca_tdf::sca_signal<double> sig_q_lp;
  sca_tdf::sca_signal<double> sig_i_rounded;
  sca_tdf::sca_signal<double> sig_q_rounded;
  sca_tdf::sca_signal<double> sig_re_real[N];
  sca_tdf::sca_signal<double> sig_re_imag[N];
  sca_tdf::sca_signal<double> sig_re_out_real[N];
  sca_tdf::sca_signal<double> sig_re_out_imag[N];
  sca_tdf::sca_signal<bool> sig_demapper[N];
  sca_tdf::sca_signal<bool> sig_received;
 private:
  //module instantiation
  q_mixer_re* q_mixer_re_sub;
  lp* i_lp;
  lp* q_lp;
  downsample* i_round,
  downsample* q_round;
  s2p<double,N>* s2p_r_sub;
  s2p<double,N>* s2p_i_sub;
  fft_ifft<N>* fft_sub;
  qam_demap* qam_demapper_sub[N];
  p2s<bool,N>* p2s_sub;
```

Listing 3.5 OFDM receiver(2)

```
public:
  //Constructor
  ofdm_re(sc_core::sc_module_name n,double mixer_fc,int
      demap_p,double bit_f,int dout_rate,double _amp=1, int
      p2s_rate=1)
    {

      int mixer_rate=(int)floor(dout_rate*log2(demap_p)*
          mixer_fc/bit_f);
      q_mixer_re_sub = new q_mixer_re("mixer_sub",mixer_fc,
          _amp,mixer_rate,false,0.,0.);
      q_mixer_re_sub->in(in);
      q_mixer_re_sub->i_out(sig_in_i);
      q_mixer_re_sub->q_out(sig_in_q);

      i_lp = new lp("i_lp",mixer_fc/1000.0);
      i_lp->in(sig_in_i);
      i_lp->out(sig_i_lp);

      q_lp = new lp("q_lp",mixer_fc/1000.0);
      q_lp->in(sig_in_q);
      q_lp->out(sig_q_lp);

      i_round = new downsample("i_round", mixer_rate,
          mixer_rate);
      i_round->in(sig_i_lp);
      i_round->out(sig_i_rounded);

      q_round = new downsample("q_round", mixer_rate,
          mixer_rate);
      q_round->in(sig_q_lp);
      q_round->out(sig_q_rounded);

      s2p_r_sub = new s2p<double,N>("s2p_r_sub",1);
      s2p_r_sub->in(sig_i_rounded);
      for(int i=0;i<N;i++)
        {
          s2p_r_sub->out[i](sig_re_real[i]);
        }

      s2p_i_sub = new s2p<double,N>("s2p_i_sub",1);
      s2p_i_sub->in(sig_q_rounded);
      for(int i=0;i<N;i++)
        {
          s2p_i_sub->out[i](sig_re_imag[i]);
        }
```

Listing 3.6 OFDM receiver(3)

```
        fft_sub = new fft_ifft<N>("fft_sub","FFT");
        for(int i=0;i<N;i++)
          {
            fft_sub->in_real[i](sig_re_real[i]);
            fft_sub->in_imag[i](sig_re_imag[i]);
            fft_sub->out_real[i](sig_re_out_real[i]);
            fft_sub->out_imag[i](sig_re_out_imag[i]);
          }

        for(int i=0;i<N;i++)
          {
            sc_core::sc_module_name nm(((string)"
                qam_demapper_sub_"+(char)(i+65)).c_str());
            qam_demapper_sub[i] = new qam_demap(nm,demap_p);
            qam_demapper_sub[i]->out(sig_demapper[i]);
            qam_demapper_sub[i]->in_i(sig_re_out_real[i]);
            qam_demapper_sub[i]->in_q(sig_re_out_imag[i]);
          }

        p2s_sub = new p2s<bool,N>("p2s_sub",p2s_rate);
        for(int i=0;i<N;i++)
          {
            p2s_sub->in[i](sig_demapper[i]);
          }
        p2s_sub->out(out);
    }
};
```

Listing 3.7 Parallel to serial converter

```
/*************************************** parallel to serial
     converter *********************************/

template <class T, int N>

  SCA_TDF_MODULE(p2s) {

  sca_tdf::sca_in<T> in[N];                        // input
       ports
  sca_tdf::sca_out<T> out;                         // output
       port

 private:
  int in_rate;
  int out_rate;

  void set_attributes()
  {
    for (int i=0;i<N;i++)
      {
        in[i].set_rate(in_rate);
      }
    out_rate=in_rate*N;
    out.set_rate(out_rate);
  }

  void processing()
  {
    T data[out_rate];
    int k=0;
    for (int i = 0; i < N; i++)
      {
        for(int j=0;j<in_rate;j++)
          {
            data[k]=in[i].read(j);
            k++;
          }
      }

    for(int j=0;j<out_rate;j++)
      {
        out.write(data[j],j);
      }
  }

 public:
  p2s(sc_core::sc_module_name n, int _in_rate=1)
    {
      in_rate=_in_rate;
    }
};
```

Listing 3.8 Serial to parallel

```
/************************************serial to paralle
    converter***************************************/

template <class T,int N>

  SCA_TDF_MODULE(s2p) {

  sca_tdf::sca_in<T> in;                    // input
  sca_tdf::sca_out<T> out[N];     // output

 private:
  int out_rate;
  int in_rate;

  void initialize(){};

  void set_attributes()
  {
    for (int i=0;i<N;i++)
      {
        out[i].set_rate(out_rate);
      }
    in_rate=N*out_rate;
    in.set_rate(in_rate);
  }

  void processing()
  {
    T symbol[N*out_rate];
    for (int i = 0; i < in_rate ; i++)
      {
        symbol[i]=in.read(i);
      }
    int k=0;
    for (int i=0;i<N;i++)
      {
        for(int j=0;j<out_rate;j++)
          {
            out[i].write(symbol[k],j);
            k++;
          }
      }
  }
 public:
  s2p(sc_core::sc_module_name n, int _out_rate=1)  {
    out_rate=_out_rate;
  }
};
```

Listing 3.9 IFFT(1)

```
/*************************************** FFT/IFFT
    **************************************************************
    */

template <int N>

SCA_TDF_MODULE(fft_ifft) {

 public:
  sca_tdf::sca_in<double> in_real[N];
  sca_tdf::sca_in<double> in_imag[N];

  sca_tdf::sca_out<double> out_real[N];
  sca_tdf::sca_out<double> out_imag[N];

 private:
  string mode;
  int isign;
  double ZERO_THRESHOLD;
  bool debug;

  void set_attributes()
  {
  }

  void initialize()
  {
  }

  void processing() {

    vector<double> data;
    double wtemp,wr,wpr,wpi,wi,theta;
    float tempr,tempi;
    int istep;
    debug=false;

    for (int j=0;j<N;j++)
      {
        data.push_back(in_real[j].read());
        data.push_back(in_imag[j].read());
      }


    int data_size=data.size();
    if(debug==true)
      {
        cout<< "before rearrange" <<endl;
        for(int j=0;j< data_size;j++)
          {
            cout<< data.at(j) <<endl;
          }
      }
```

Listing 3.10 IFFT(2)

```
    int j=1;                                                    //
        This is the bit-reversal section of the routine.

     for (int i=1; i < data_size; i+=2)
       {
         if (j > i) {
           swap(data.at(j-1), data.at(i-1));
           swap(data.at(j), data.at(i));
         }
         int m = data_size >> 1;
         while (m>=2 && j>m) {
           j -= m;
           m >>= 1;
         }
         j += m;
       }
     if(debug==true)
       {
         cout<< "after rearrange" <<endl;
         for(int j=0;j< data_size;j++)
           {
             cout<<data.at(j)<<endl;
           }
       }
     int mmax=2;

     while (data_size>mmax)
       {
         istep = mmax << 1;
         //      theta = isign*(6.28318530717959/mmax);
         theta = isign*(2*M_PI/mmax);
         wtemp = sin(0.5*theta);
         wpr = -2.0*wtemp*wtemp;
         wpi = sin(theta);
         wr = 1.0;
         wi = 0.0;
         for (int m=1; m<mmax; m+=2)
           {
             for (int i=m; i<=data_size; i+=istep)
               {
                 int j = i + mmax;
                 tempr = wr*data[j-1]-wi*data[j];
                 tempi = wr*data[j]+wi*data[j-1];
                 data[j-1] = data[i-1]-tempr;
                 data[j] = data[i]-tempi;
                 data[i-1] += tempr;
                 data[i] += tempi;
               }
             wr = (wtemp=wr)*wpr - wi*wpi + wr;
             wi = wi*wpr + wtemp*wpi + wi;
           }
         mmax = istep;
       }
```

Listing 3.11 IFFT(3)

```
          if(debug==true)
          {
            cout<< "after transform" <<endl;
            for(int j=0;j< data_size;j++)
              {
                cout<<data.at(j)<<endl;
              }
          }

        if (mode=="IFFT")                                    //devide by N if IFFT
          {
            for(int j=0;j< data_size;j++)
              data[j]=data[j]/N;
          }
        if(debug==true)
          {
            cout<< "divided by N" <<endl;
            for(int j=0;j< data_size;j++)
              {
                cout<<data.at(j)<<endl;
              }
          }

        int k=0;                                             // output
        for (int i=0;i<N;i++)
          {
            out_real[i].write(data.at(k));
            out_imag[i].write(data.at(k+1));
            k+=2;
          }
      }

  public:
    fft_ifft(sc_core::sc_module_name n, string _mode)
      {
        mode=_mode;
        if (mode=="FFT")
          isign=-1;
        else if(mode=="IFFT")
          isign=1;
        else
          cout<<"Error:mode can only be FFT or IFFT"<<endl;
      }
};
```

Listing 3.12 Downsampler(1)

```
/************************Down Sampler
    ************************ */
// This module decreases the data rate of the input signal.
    It reads _rate values and writes one value to the output.
    With the parameter _sel it is possible to selcet the
    _selth data which you want to output. It is an integer
    number between 1 and _rate.
SCA_TDF_MODULE(downsample) {

  sca_tdf::sca_in<double>  in    ;
  sca_tdf::sca_out<double>  out   ;

 private:
  int rate ;

  int sel;

  void set_attributes() ;

  void initialize() ;

  void processing();

 public:
  // The constructor takes the value of parameter _sel and
      the data rate of the input port.They are set to 1 and 1
      by default, respectively.
  downsample(sc_core::sc_module_name nm,int _sel=1,int _rate
      =1);
};
```

Listing 3.13 Downsampler(2)

```
/***************************Down Sampler
    ****************************************************/


    void downsample::set_attributes() {

      in.set_rate(rate);

    }

    void downsample::initialize() {}

    void downsample::processing() {

      out.write(in.read(sel-1)) ;

    }

    downsample::downsample(sc_core::sc_module_name nm, int
        _sel, int _rate_in) {

      rate= _rate_in ;
      sel=_sel;
    }
```

Listing 3.14 Parameters for our simulation

```
// This file contains the main declararations


// Bit rate of the digital signal sourece, 1 bit per 1/freq_byte second;
double freq_bit = 1000000.;

// Frequency of carrier wave in Hz
double freq_carrier = 1000000000.0;

// Cutoff-frequency of the lowpass-filter in the receiver.
double freq_cutoff = freq_carrier/1000;

// Note that this datarate can basically be set independant of the frequencies byteFreq
// and Carrier Frequency, but a value too low would result in an undesirable outcome
// regarding Nyquist's sampling theorem.By the  formula above, about 30 token
// per sine period are ensured
int data_rate=(int)floor(30.*4*freq_carrier/freq_bit);
```

3.4.5 OFDM Transceiver with Library Usage

In this section the building block library is used to implement the transceiver system, immensely reducing the implementation effort. The essential parts of a complete implementation can be seen in Section 3.4.4. Since the building block library provides modules with high abstraction level, the implementation follows the basic block structure shown in Fig. 3.17.

3.4.6 Advantage of High-Level Signal Processing with the BBL

The usage of the library speeds up the initial process of rapid prototyping (RAD) a system and eases the initial phase of design space exploration. This is achieved by providing a lot of modules for special purposes. The main advantage of the library lies in the fact that the modules are not necessarily monolithic; the more powerful ones are assembled from smaller building block modules, which are available for the system designer too, so it is easy to build upon the modules and the building block library can be used itself as a source of information on how to use or build generally usable modules with SystemC AMS for simulating analog system parts in an efficient way.

An example for such a non-monolithic module is, for instance, the `ofdm_se` module in the example in Section 3.4.5. The OFDM sender module internally reuses the modules `s2p` (serial to parallel converter), `qam_map` (a quadrature amplitude

Listing 3.15 Implementation: boilerplate code, variable declarations and handling input for parameters

```
#include "systemc-ams.h"
#include "ofdm_declarationslib.h"
#include <bb/bb.h>

using namespace ahes::bb;

SCA_TDF_MODULE(drain) {
  sca_tdf::sca_in<bool> in;
  SCA_CTOR(drain) {}
};

int sc_main(int argc, char* argv[])
{
  sc_set_time_resolution(0.01, SC_NS);  // setting the time
      resolution


  /* ********* defining signals and parameters *********** */

  sca_tdf::sca_signal<bool>   sig_stimuli;
  sca_tdf::sca_signal<double> sig_out;
  sca_tdf::sca_signal<double> noise_out;
  sca_tdf::sca_signal<double> sig_noise;
  sca_tdf::sca_signal<bool>   sig_received;

  int constl_dim;
  double ampl_se;
  double ampl_re;
  double n_va;
  double n_mean;
  double attent;

  /* ******** setting parameters for simulation ********** */

  cout << "constel_dim = "; cin >> constl_dim;
  cout << "ampl_se = "; cin >> ampl_se;
  cout << "ampl_re = "; cin >> ampl_re;
  cout << "n_va = "; cin >> n_va;
  cout << "n_mean = "; cin >> n_mean;
  cout << "attent = "; cin >> attent;
```

Listing 3.16 Implementation: transceiver itself

```
/* ********* instantiating TDF-modules ********* */

// Stimuli generator: As a signal source it generates uniformly
// distributed random sequence of bits on its output."sig_stimuli"
// is the corresponding stimul signal.
rand_bool i_stimuli("stimuli",1);
i_stimuli.out(sig_stimuli);
i_stimuli.out.set_timestep(1/freq_bit,SC_SEC);

// OFDM Transmitter: It takes the sequence of bits from the
// stimuli generator and converts them to radio frequency signal
// which will be sent per an antenna.
ofdm_se<8> i_tran("transmitter",constl_dim,freq_carrier,data_rate,ampl_se);
i_tran.in(sig_stimuli);
// "sig_out" is the radio frequency signal which will be sent per antenna.
i_tran.out(sig_out);

// This part models the environment. A noise generator generates
// white gaussian noises and adds them to the output signal of
// the Transmitter. Modeling of attentuation of the environment is also allowed.
air i_air("air",attent,"gauss_white",n_va,n_mean,data_rate);
i_air.in(sig_out);
i_air.out(sig_noise);

// OFDM Receiver: It takes the signal from the environment and
// translats them back to the original sent digital signal.
ofdm_re<8> i_receiver("receiver",constl_dim,freq_carrier,data_rate,freq_cutoff,ampl_re);
i_receiver.in(sig_noise);
i_receiver.out(sig_received);

// This is only a module used to consume tokens as we can not let a
// systemc-ams scheduling loop open.
drain drn("drn");
drn.in(sig_received);
```

Listing 3.17 Implementation: boilerplate for signal tracing and simulation start

```
    /* ******* tracing signals */

    // VCD trace file: With the VCD trace file, one can check
        if the received
    // signal and stimuli have identical values.
    sca_util::sca_trace_file* wave = sca_util::
        sca_create_vcd_trace_file("wave");

    sca_util::sca_trace(wave, sig_stimuli,   "stimuli");
    sca_util::sca_trace(wave, sig_out,       "sig_wo_noise");
    sca_util::sca_trace(wave, sig_received, "sig_received");
    sca_util::sca_trace(wave, sig_noise,     "sig_wt_noise");

    // Simulation time: 0.05ms
    sc_start(0.05, SC_MS);
    sca_util::sca_close_vcd_trace_file( wave );

    return 0;
}
```

modification mapper),`fft_ifft` (an inverse fast Fourier transformation module), `p2s` (a parallel to serial converter) and `q_mixer_tr`.

References

1. ANDRES Project Homepage 2009. http://www.andres.offis.de/. Accessed on September 20, 2010.
2. Cai L, Gajski D (2003) Transaction level modeling in system level design. In: Technical Report 03-10, Center for Embedded Computer Systems, University of California, Irvine
3. Herrholz A, Oppenheimer F, Schallenberg A, Nebel W, Grimm C, Damm M, Herrera F, E Villar AMF, Martinez M (2007) Andres-analysis and design of run-time reconfigurable, heterogeneous systems. In: Workshop on "Adaptive Heterogeneous Systems-On-Chip and European Dimensions" in the Design Automation and Test in Europe 2007, DATE'07, Nice, France. DATE 07 Friday Workshop Notes, 64–71.
4. Mahlknecht, Durante MS (2009) An ultra low power wakeup receiver for wireless sensor nodes. In: Proceedings of the Third International Conference on Sensor Technologies and Applications (SENSORCOMM 2009), Athens, Greece, 2009.
5. OSCI (2008) OSCI TLM2.0 draft2. Open SystemC Initiative, http://www.systemc.org. Accessed on September 20, 2010.
6. OSCI (2010) SystemC^{TM}. Open SystemC Initiative, http://www.systemc.org. Accessed on September 20, 2010.
7. Vachoux A, Grimm C, Einwich K (2005) Extending SystemC to support mixed discrete-continuous system modeling and simulation. In: IEEE Symposium on Circuits and Systems
8. OMNeT++ Project Homepage, Accessed on September 2010.

Chapter 4
Remote Sensing of Car Tire Pressure

Thomas Herndl

Abstract State-of-the-art tire pressure monitoring systems (TPMS) are wireless sensor nodes mounted on the rim. Attaching the node on the inner liner of a tire allows sensing of important additional technical parameters, which may be used for improved tracking and engine control, feedback to the power train and car-to-car communication purposes. Thus a significant step in car control appears feasible. Those new features come at a price: the maximum weight of the sensor is limited to 5 g including package, power supply, and antenna. Robustness is required against extreme levels of acceleration of up to 3,000 g ($g = 9.81\,\mathrm{m/s^2}$). The node size is limited to about 1 cm^3 to avoid high force gradients due to device deformation and finally, a 10 year power supply lifetime must be achieved. In this chapter we present a self-sufficient tire-mounted wireless sensor node.

- with a bulk acoustic wave (BAW)-based low-power FSK transceiver;
- pioneered for an energy scavenger-based low-volume and low-weight power supply; and
- a 3D vertical chip stack for best compactness, lowest volume, and highest robustness for pressure, inertia, and temperature sensing.

Keywords TPMS · Tire pressure monitoring system · Energy harvesting · Energy scavenging · Wireless sensor network · 3D system integration · Sensor system

4.1 Motivation for Tire Pressure Monitoring Systems

More than 20 years have passed since in 1986 the Porsche 959 car was first equipped with a built-in tire pressure monitoring system (TPMS) [1] and since then millions of cars have followed, because of the large variety of safety risks surrounding incorrectly inflated or damaged tires. In fact, underinflated tires are more prone to stress damage, have less lateral traction, a shorter tread life, and are more vulnerable to flat tires and blowouts (Fig. 4.1).

T. Herndl (✉)
Infineon Technologies Austria AG, Vienna, Austria
e-mail: thomas.herndl@infineon.com

T.J. Kaźmierski, S. Beeby (eds.), *Energy Harvesting Systems*,
DOI 10.1007/978-1-4419-7566-9_4, © Springer Science+Business Media, LLC 2011

Fig. 4.1 Tire blowout

Furthermore, underinflated tires can increase the distance required for a vehicle to stop, particularly at wet surface conditions. Investigations in the USA have shown that

1. one-fifth of all tires are up to 40% under their correct pressure;
2. 10% drop in pressure cuts a tire's service life by 15%;
3. each 0.2 bar drop in pressure increases fuel consumption by 1.5%;
4. 75% of flat tires due to insufficient pressure or gradual pressure loss;
5. tire problems are third most common cause of vehicle breakdowns; and
6. ~250,000 accidents a year (the USA) can be traced back to insufficient tire pressure.

Thus, triggered by the US National Highway Traffic Safety Administration (NHTSA) recent laws have put the spotlight on the issue of tire pressure monitoring. The 2000 TREAD (Transportation Recall Enhancement, Accountability, and Documentation) Act requires automobile manufacturers to gradually provide TPMS for all cars sold in the USA [2, 3]. As minimum requirements a "low tire pressure" indication lamp and a "malfunction" indication lamp have been defined (Fig. 4.2).

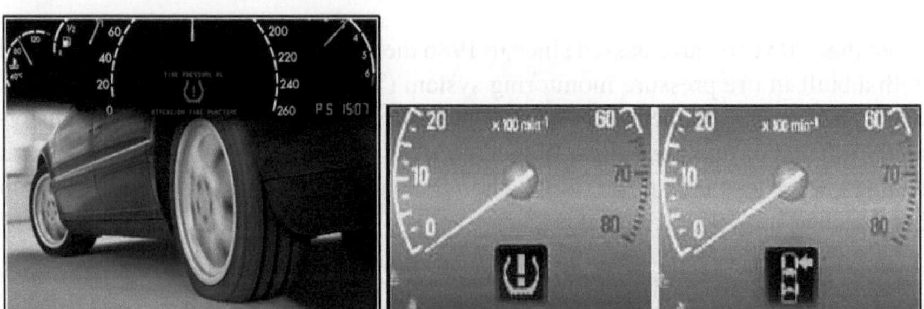

Fig. 4.2 Obligatory in the USA: low tire pressure indication and malfunction indicators

While for the time being in Europe TPMS is rather considered as non-mandatory comfort feature, it is very likely that automotive manufacturers will be obliged by law to equip new cars with TPMS, starting from 2013.

4.2 Overview of State-of-the-Art TPMS

State-of-the-art direct tire pressure monitoring systems (TPMS) are wireless sensor nodes mounted on the rim. Basically, there are mainly three technologies competing for their market presence.

4.2.1 Direct Battery-Based (Active) Systems

Direct battery-based (active) systems have their own independent power source integrated in the modules and each module contains a pressure sensor. The modules are mounted on the rim, as part of the valve (Fig. 4.3). Core functions (sensors and wireless transmitter) are integrated into a system in package (SiP). In order to fulfill lifetime requirements, the battery size is quite large and dominates the total weight and size of the TPMS module.

Only a single, central radio-receiver is required (Fig. 4.4), which is typically located in the dashboard. All in all this keeps the total system costs moderate. Due

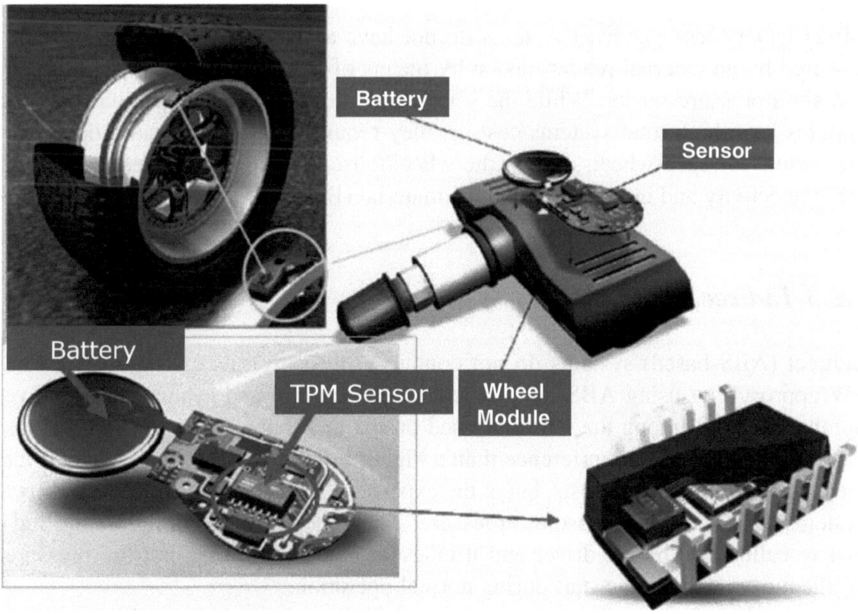

Fig. 4.3 Rim-mounted tire pressure monitoring system and subcomponents (state of the art)

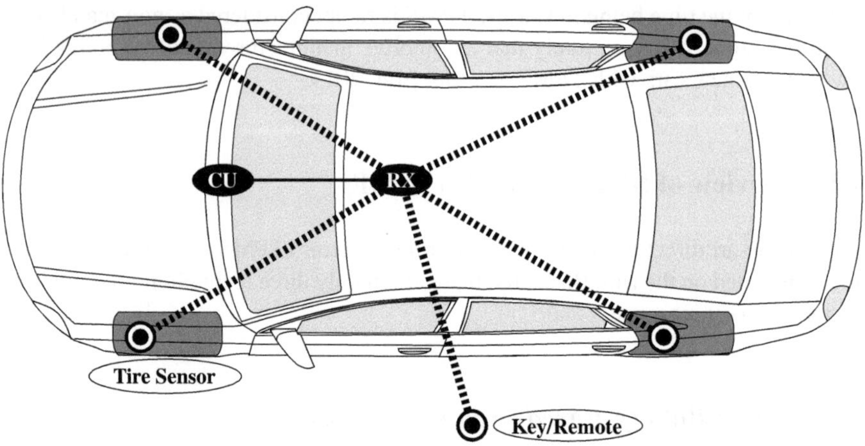

Fig. 4.4 Tire pressure monitoring system communicating with ECU/dashboard in a star network topology

to their superior performance figures at moderate costs, direct battery-based systems constitute far the major share among all available technologies for tire pressure monitoring.

4.2.2 Direct Battery-Less (Passive) Systems

Direct battery-less (passive) systems do not have a power source, rather they are powered by an external reader device by means of an electromagnetic field. They contain a pressure sensor. While the sensor modules are cheap, direct battery-less systems have high total systems cost, as they require a powering/reader device in the vicinity of each wheel, e.g., in the wheelhouse, which complicates the wiring for connectivity and increases the bill of material (BOM).

4.2.3 Indirect (ABS-Based) Systems

Indirect (ABS-based) systems do not contain a pressure sensor. They rely on an SW approach by using ABS data (wheel speed sensors) and evaluate the relative number of turns among the wheels, based on the fact that a lower inflated tire has a smaller roll radius/circumference than a higher inflated one. Logically, they offer the cheapest total system costs, but at the expense of accuracy and comfort: Indirect systems cannot measure absolute pressure. They need initial calibration and regular re-calibration by the driver and it takes several minutes for first tire pressure notifications upon engine start during normal operation.

Due to the position of the sensor node in the valve, the node has only very loose contact to the road surface, which prevents additional important parameters from being monitored. Hence, state-of-the-art rim-mounted direct systems, equipped in many (US) cars, have limited sensing capabilities, restricted to movement detection, tire pressure and tire temperature monitoring. Further on, the rim-mounted TPMS node "does not intrinsically know" which tire it carries, hence it is not possible to trustworthily report even "simple" additional tire-specific static information, like a tire-ID reflecting the mounted tire type (summer/snow tires; spikes), which would be a very valuable parameter for the ABS system to further optimize the stopping distance of the vehicle in emergency situations. Indeed, additional parameters can be reported when moving the TPMS module to within the tire, as shown in Fig. 4.5.

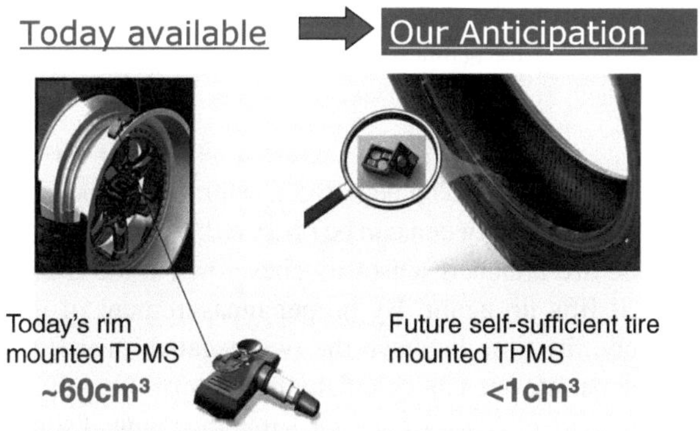

Fig. 4.5 Rim-mounted versus in-tire-mounted TPMS module

4.3 Motivation and Requirements for In-Tire TPMS

Attaching the TPMS node on the inner liner of a tire allows the extraction of additional information for comfort and safety beyond tire pressure and temperature, such as

- vehicle load;
- wheel speed;
- wear;
- tire friction, side slip, road condition; and
- integrated tire-ID for logistic purposes (fleet management, localization, automatic detection of tire change, and tire type for input to ABS).

They may be used for improved tracking and engine control, feedback to the power train and car-to-car communication purposes and can therefore contribute to further enhance driving safety.

Technically, additional parameters can be derived from acceleration signals occurring at the border between the road surface and the tire. As can be seen from Fig. 4.6, which shows signal traces measured with an acceleration sensor attached

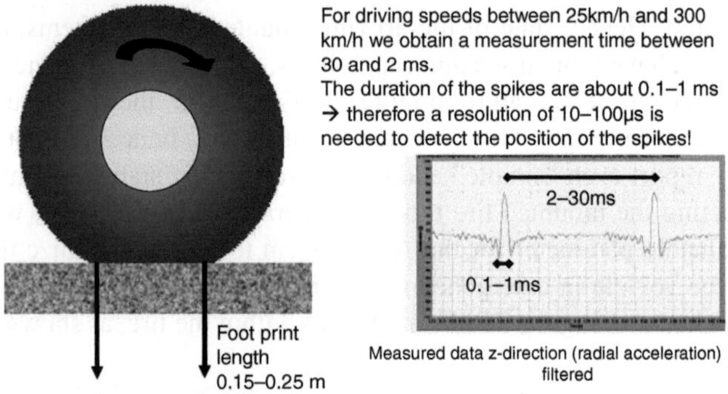

Fig. 4.6 Technical background for additional in-tire parameter monitoring: acceleration signal trace from the tire

at the inner liner of the tire, two steep negative acceleration peaks and one dominant positive acceleration peak in between can be observed: the first negative peak occurs at positions when the tire-attached sensor touches down at the road surface and the second peak when it lifts up again. By proper measurement of the positive peak duration (respectively, the time between the two negative peaks) and the distance between the positive peaks, one can calculate the footprint length and other important parameters listed above. Additionally the same mechanical shock signal can be used as the source for a vibration harvesting unit, translating mechanical power into electric power for supplying a self-sufficient, battery-less sensor node.

Those new features come at a price: The maximum weight of the in-tire sensor is limited to $\sim$5 g including package, power supply, and antenna. Robustness is required against extreme levels of acceleration of up to 3,000 g ($g = 9.81\,\mathrm{m/s^2}$). The node size is limited to about 1 $\mathrm{cm^3}$ to avoid high force gradients due to device deformation and finally, a 5- to 10-year power supply lifetime must be achieved.

4.4 A Self-Sufficient In-Tire TPMS Demonstrator

In this chapter we present a self-sufficient tire-mounted wireless sensor node. Figure 4.7 shows a block diagram of the in-tire TPMS demonstrator, consisting of a MEMS sensor, a power supply module, a microcontroller ASIC, and a transceiver ASIC which directly generates the RF carrier by using a BAW resonator.

The key innovations of the demonstrator are

- a bulk acoustic wave (BAW)-based frequency shift keying (FSK) transceiver;
- pioneered for an energy scavenger-based low-volume and low-weight power supply;
- a 3D vertical chip stack for best compactness, lowest volume, and highest robustness; and
- molded interconnect device (MID) package solution with integrated antenna

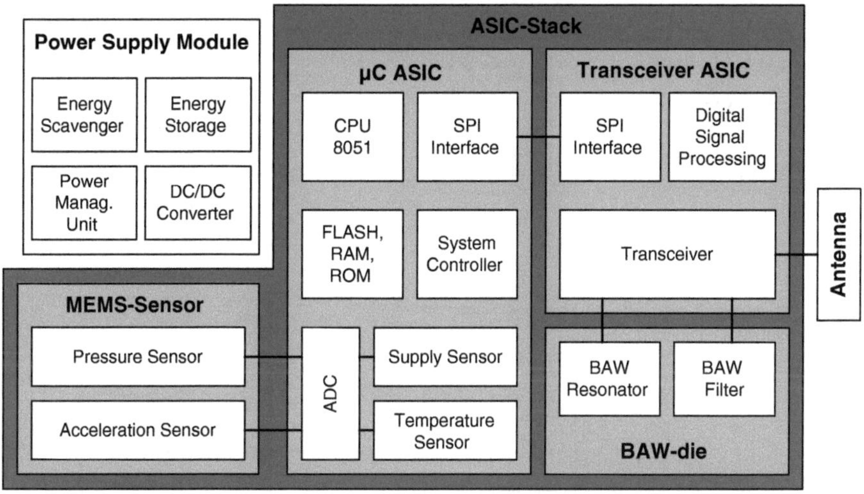

Fig. 4.7 Block diagram of the in-tire TPMS sensor node demonstrator

for pressure, inertia, and temperature sensing and will be presented in more detail in the following.

4.4.1 RF Transceiver

For use in the proposed sensor node a BAW based RF transceiver has been developed (Fig. 4.8), which directly generates a 2.4 (2.11) GHz carrier (two variants have been designed, operating at carrier frequencies of 2.11 and 2.4 GHz) by using a BAW resonator with a size of only 0.02 mm^2. It avoids the employment of a bulky and shock-sensitive crystal and a phase-locked loop (PLL), which makes the system more robust and radically reduces the turn-on time to a few microseconds from several milliseconds as in state-of-the-art crystal oscillator systems. This improves the overall power consumption [1, 4].

For the receiving section an image-reject architecture has been chosen with BAW resonators integrated into the LNA for filtering [5]. Typically the parasitic capacitances of the electrodes of the BAWs are not equal, since the bottom electrodes have a higher capacitance against substrate. Therefore it is advantageous to use two BAWs, one in each branch of the differential LNA. Since the attenuation near the series resonance frequency of the BAW is very high, low side injection of the local oscillator (LO) signal leads to a high additional suppression of the image frequency, allowing for a low intermediate frequency (IF) of 10.7 MHz. In contrast to [6], the presented receiver utilizes the narrow bandwidth of a single BAW resonator instead of a filter consisting of several resonators. The LNA is followed by an RC polyphase network to generate the I- and Q-phases for the image-reject mixer. At the IF the received signal is filtered and fed into a limiting amplifier delivering a binary signal,

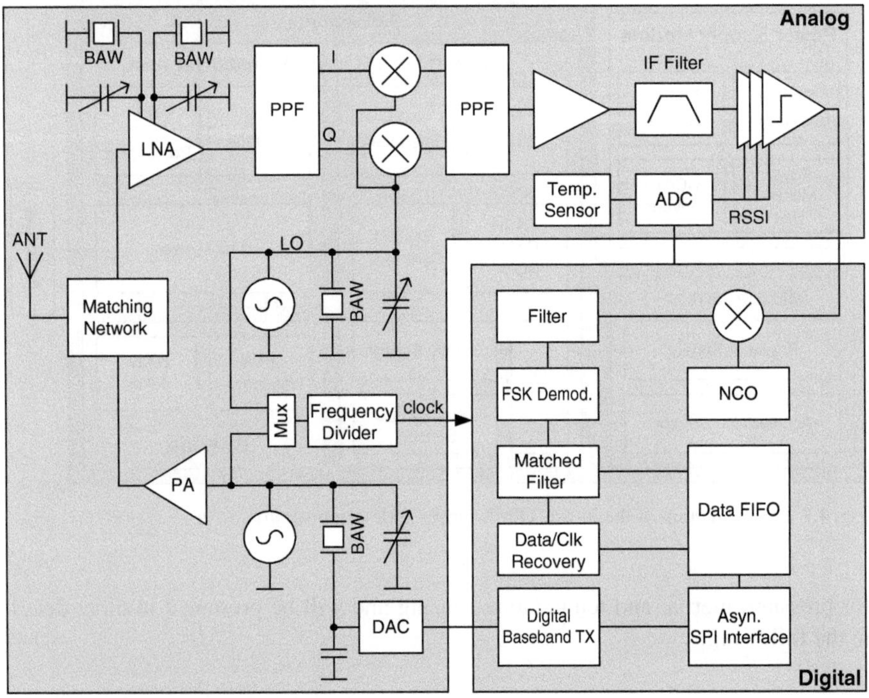

Fig. 4.8 Block diagram of the BAW transceiver

which is directly mixed into complex baseband in the digital domain. After digital filtering, demodulation, and de-framing the received payload is stored in a FIFO.

Small-sized shock-resistant BAW resonators are utilized as frequency reference for the TRX and channel selection filter in the receive chain [7]. A BAW resonator can be considered as a thin film of piezoelectric material sandwiched between two metal electrodes. When an electric field is applied between the electrodes, the structure is mechanically deformed by way of inverse piezoelectric effect and acoustic waves are launched into the bulk of the device [8].

Basically two approaches can be distinguished to prevent that the resonant energy is not absorbed by the carrier substrate. In the first one, micromachining is used to generate a membrane. The second approach is shown in Fig. 4.9, where the waves are reflected on an acoustic mirror and therefore cannot propagate any further into the substrate [8]. The benefits offered by a BAW resonator includes extremely high mechanical robustness, higher quality factor (Q) than surface acoustic wave (SAW) devices or on-chip LC tanks, small size, and lower temperature dependency and higher ESD robustness than SAW.

One major drawback of BAW devices is their temperature dependency, which is typically in the range of $-18\,\text{ppm}/^\circ\text{C}$ (Fig. 4.10).

To overcome temperature drift effects the temperature is measured and compensated via digitally controlled capacitors (9 bit) in parallel to the resonator in

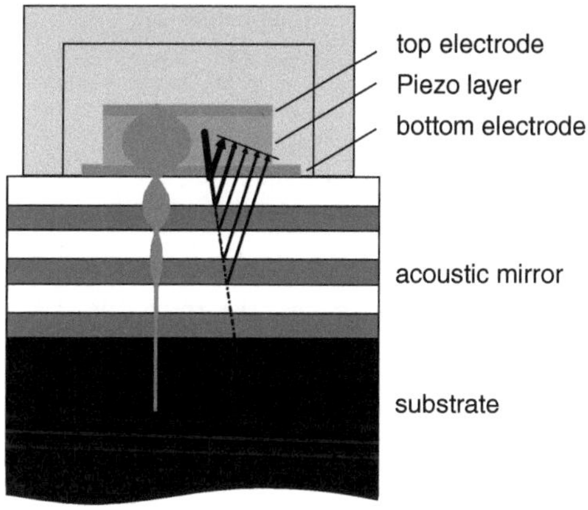

Fig. 4.9 Basic concept of a BAW resonator with acoustic mirror

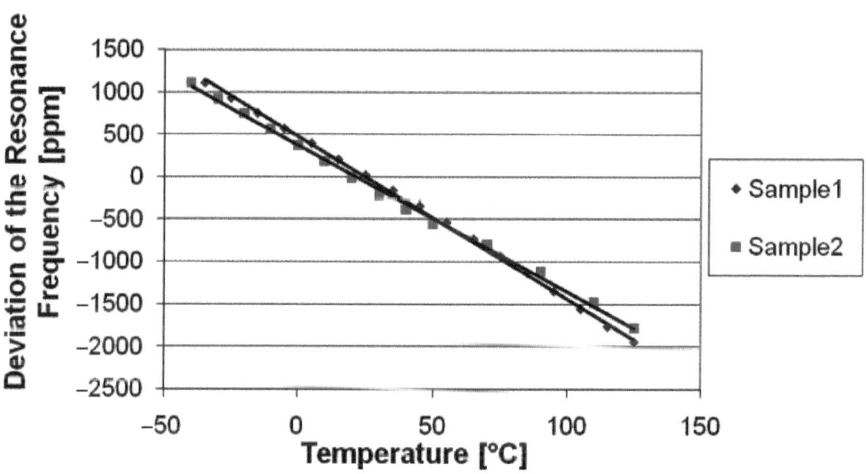

Fig. 4.10 Temperature dependency of the BAW device resonance frequency without temperature compensation

the range of −40 to +125°C. Figure 4.11 shows the resulting frequency deviation which stays within ±40 ppm for the whole temperature range after applying a proper temperature compensation algorithm. Additionally, the BAW can be tuned with a variable DC biasing, which causes a change in the stiffness of the piezoelectric BAW material. The biasing voltage causes a linear frequency shift with a slope of 40 kHz/V. Due to the fast response of the resonator, this effect is used for FSK modulation in the transmitter. Direct carrier modulation and frequency tuning are also possible with the digitally controlled tuning capacitors, but the nonlinear and process-dependent relationship between capacitance and resonance frequency is a drawback, which has to be considered when applying this method.

The transceiver ASIC was fabricated by using a 0.13 μm standard, automotive-qualified CMOS process. Its current consumption is 6 mA in transmit mode at a

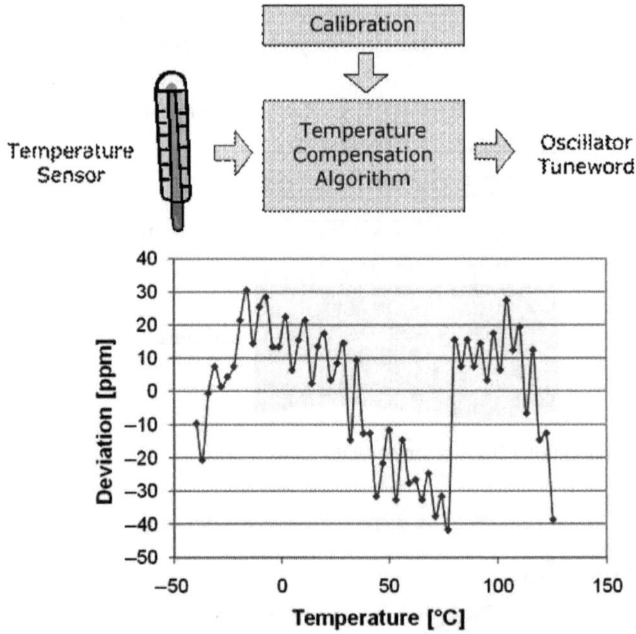

Fig. 4.11 Temperature dependency of the BAW device resonance frequency with applied temperature compensation algorithm

transmit output power of 1 d Bm and 8 mA in receive mode with a sensitivity of −90 dBm at a data rate of 50 kb/s. The turn-on time of the transceiver is only 2 μs.

4.4.2 Power Supply Subsystem

The power supply subsystem (Fig. 4.12) consists of

- miniaturized MEMS transducer device;
- high efficient power translation ASIC (0.25 μm automotive-qualified CMOS process);
- external coil for the inductive AC/DC converter; and
- energy storage device (low-leakage capacitor).

For the MEMS vibration scavenger an electrostatic transducer device has been chosen, manufactured by using high aspect ratio micromachining (Fig. 4.13) [9].

It already integrates an electret for biasing of the transducer; hence, there is no additional external voltage source required. Due to the very small mass of the miniaturized vibration scavenger and the resulting very weak mechanical coupling, only a small portion of the available mechanical energy in the tire can be actually translated into electrical energy. In-tire measurements and electromechanical simulations have shown that the scavenging system with a vibrating mass of about 30 mm^2 is able

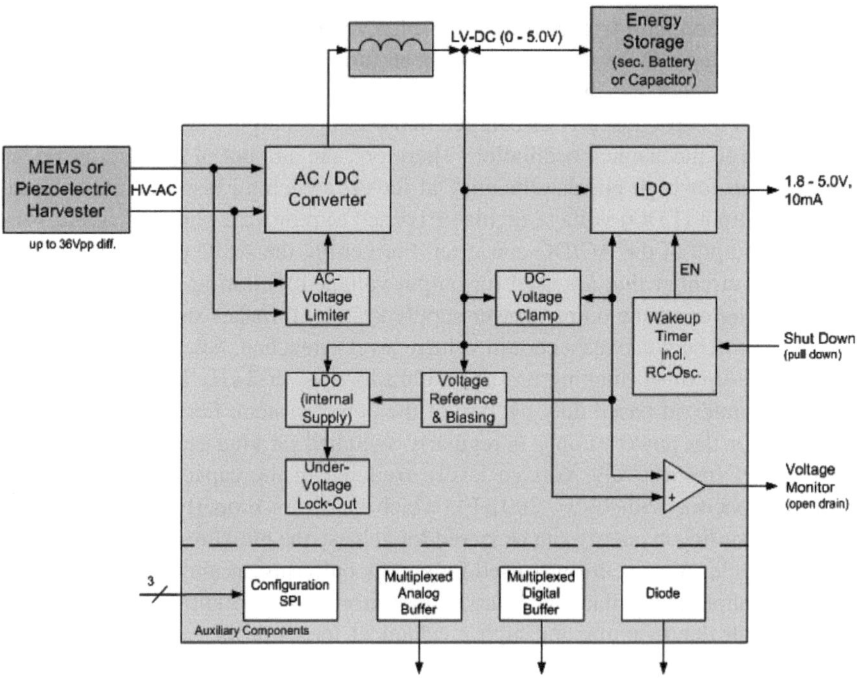

Fig. 4.12 Block diagram of the vibration scavenger-based power supply subsystem

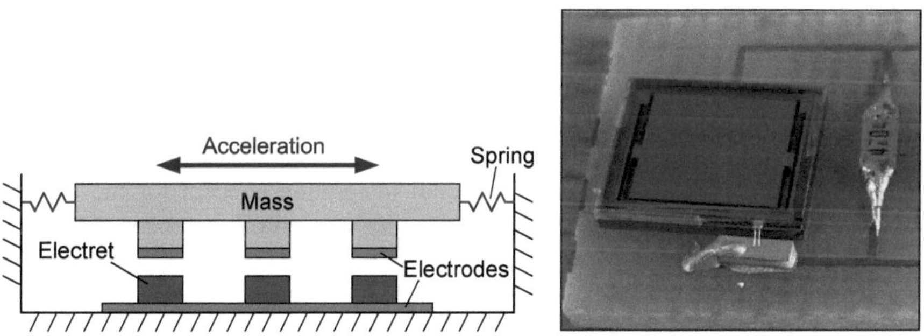

Fig. 4.13 Vibration scavenger: functional principle (*left*) and photo of the device (*right*)

to deliver some microamperes at 50 km/h driving speed, which is sufficiently high for regular reporting of the tire pressure at appropriate intervals. With increasing driving speed, the available vibration power also increases, allowing for shortening the reporting intervals, which results in higher update rates for pressure reports at higher driving speeds.

The successive power conversion ASIC integrates a high efficient inductive AC/DC down-converter for charging a capacitor up to the desired voltage level. It can process AC input voltages up to ~40 V. The input stage of the harvester interface consists of an active full bridge rectifier with peak detector. The peak detector generates the optimum trigger time instant for voltage conversion, independent of the output voltage. Once the switch has been turned on, down-conversion is done within a "single shot" via the external coil. Upon completion of the conversion cycle, the harvester's capacitance is discharged and starts recharging again with opposite polarity due to mechanical oscillation. Therefore, the amount of conversion cycles is minimized for high power efficiency at lowest switching losses. An additional low drop output (LDO) voltage regulator is used to remove a potential ripple voltage at the output of the AC/DC converter. Further on, the ASIC provides auxiliary and control circuitry that are used for output voltage monitoring and switching off and switching on of the output power supply system. It makes sure that the output voltage is switched on once a certain voltage level is reached. Alternatively the integrated wake-up timer, making use of an ultra-low power 2 kHz RC oscillator, can be used for interval-based duty cycling of the power system. For the in-tire TPMS demonstrator the power supply is regularly switched on after exceeding a certain energy level (respectively, voltage level) available in the capacitor (low-leakage ceramic capacitor with $C = 200\,\mu\mathrm{F}$), which has been properly dimensioned to ensure that sufficient energy can be stored for at least one pressure measurement and reporting cycle. A capacitor was used for energy buffering, as alternative secondary battery technologies suitable for harshest environmental conditions as present in the tire (high temperatures and high mechanical forces at the same time) are not available in the market.

4.4.3 Overall System Integration of the In-Tire TPMS Node

For demonstration purposes, both a battery-driven and a vibration harvester-driven version have been designed. In order to reach high compactness and robustness for system integration, miniaturization concepts have been applied on different levels.

Figure 4.14 shows a sketch of the overall integration concept for the battery version. For the scavenger-driven version another micro-PCB has been designed,

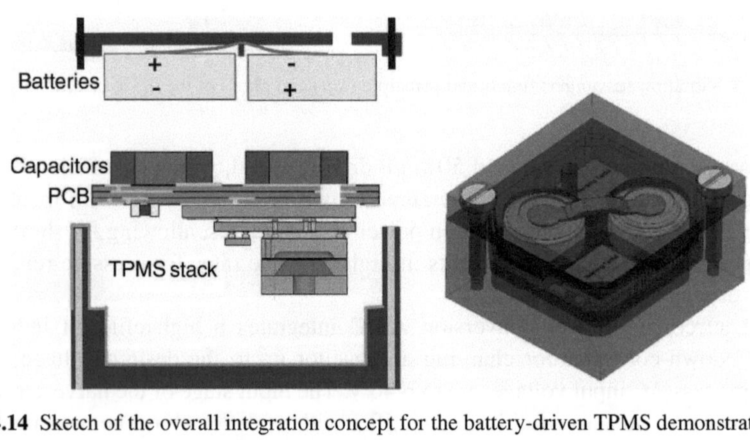

Fig. 4.14 Sketch of the overall integration concept for the battery-driven TPMS demonstrator

which replaces the lid. The space of the batteries and the capacitors is used for the placement of all subcomponents of the scavenger power supply subsystem.

All functional core components such as microcontroller, transceiver, BAW device, and sensor were co-assembled by means of 3D vertical system integration, making use of through silicon vias (TSV), applied to the RF transceiver ASIC and the sensor, and bumping technologies [10–13]. The resulting 3D sub-stack shown in Fig. 4.15 was glued and wire-bonded onto a micro-PCB (Fig. 4.16). At the opposite side of this micro-PCB tiny cell batteries and buffer capacitors were placed for the battery-driven version.

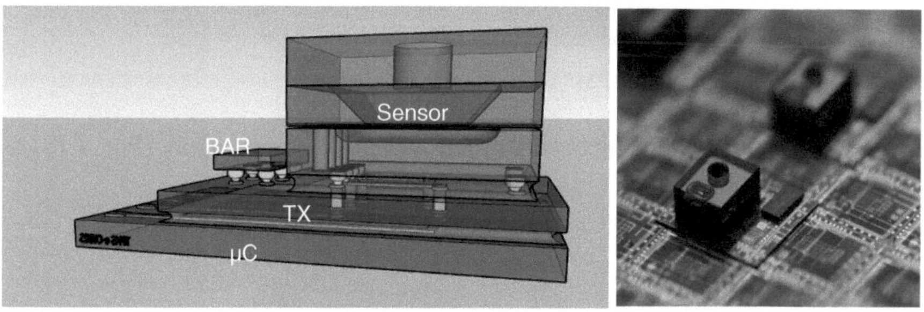

Fig. 4.15 Sketch and photo of the TPMS 3D chip sub-stack (*source*: SINTEF)

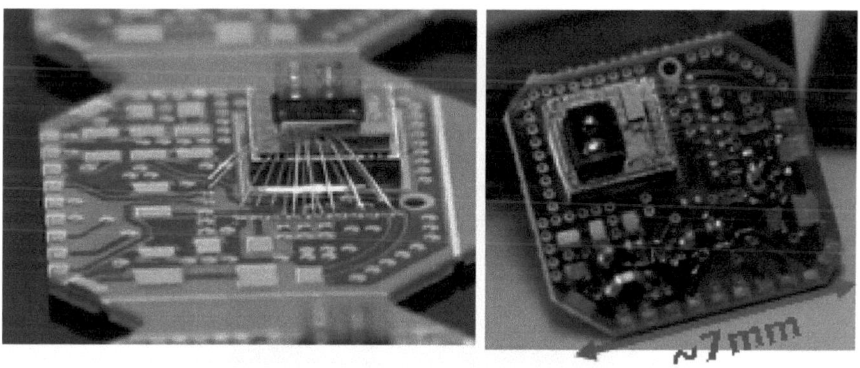

Fig. 4.16 TPMS 3D chip sub-stack assembled on the micro-PCB

For packaging, a cube-like plastic 3D substrate was designed using MID technology (molded interconnect device, Fig. 4.17), allowing for integration of conducting lines directly onto the package surface. This ability was beneficially used for shaping a loop antenna and providing the connectivity between the functional TPMS subsystem and the power supply subsystem. The overall dimensions of the MID package are only $1.2 \times 1.3 \times 0.64\,\text{cm}^3 = 0.998\,\text{cm}^3$, which is less than $1\,\text{cm}^3$.

Figure 4.17 shows the lid with metal springs used for the battery-driven version. For the scavenger-driven version a micro-PCB of the same size as the lid was

Fig. 4.17 MID package for the TPMS sensor node demonstrator

used, comprising all subcomponents of the scavenger power supply (Fig. 4.18) as described in Section 4.4.2.

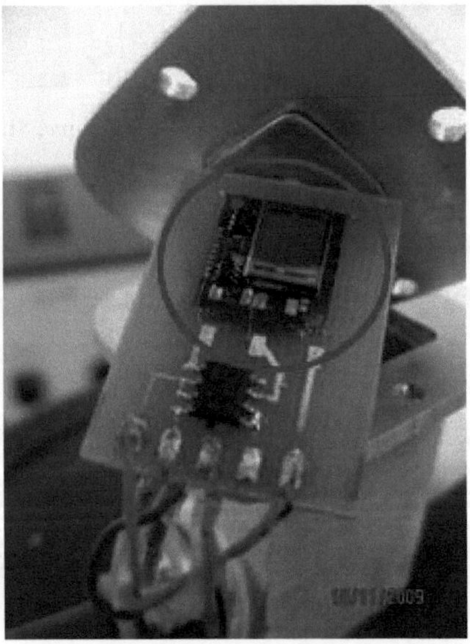

Fig. 4.18 Vibration harvester power supply board of the in-tire TPMS demonstrate

Figures 4.19 and 4.20 finally present different views of the fully assembled TPMS demonstrator node (battery version and harvester version, respectively); the hole in the MID package serves as pressure inlet. The transparent package shown in Fig. 4.19 has been designed for visualization purposes only and is not functional.

Based on the fully assembled TPMS demonstrators, system characterization has been performed. One key performance parameter is the overall power consumption, respectively, energy consumption. Figure 4.21 shows the measured current profile for a system cycle (pressure measurement and reporting). Three

Fig. 4.19 Fully assembled TPMS demonstrator (battery version)

Fig. 4.20 Fully assembled TPMS demonstrator (vibration scavenger version)

phases can be observed: Upon activation of a system cycle, the pressure value is sensed and post-processed by the microcontroller. After completion of the sensor measurement phase the system is configured for the following data transmission. A temperature measurement is done and temperature compensation of the RF frequency is performed (see also Fig. 4.11). Further on, the RF transmitter is properly configured. Finally, the data are transmitted over the wireless channel, which terminates the reporting cycle. This is done in a cyclic manner.

One whole cycle requires about 5 ms. The average current consumption over the active cycle is about 5 mA. When activating the cycle every 10 s, the average current consumption significantly drops down to about 3 μA, which can be delivered by the vibration scavenging power supply system at driving speeds of $\sim$50 km/h. When driving at lower speeds, the reporting interval needs to be prolonged, as it takes longer to charge the storage capacitor due to the diminished electrical power from the harvester.

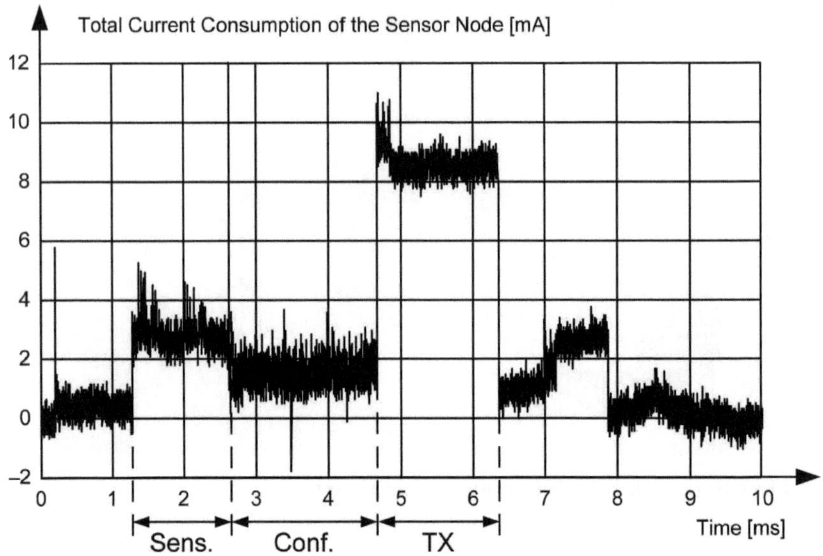

Fig. 4.21 Current consumption profile of the in-tire TPMS demonstrator for a measurement and data reporting cycle

Technology	
Transceiver	CMOS, 0.13μm
Microcontroller	CMOS, 0.24μm
Current Consumption	
Oscillator core + biasing	900uA
Transmitter, 1dBm	6mA
Receiver, -90dBm, 50kBit/s	8mA
Overall peak current	<10mA

Fig. 4.22 Key performance figures of the in-tire TPMS demonstrator

Figure 4.22 summarizes key performance figures of the in-tire TPMS demonstrator. It shall be emphasized that at no time the overall current consumption exceeds 10 mA, which relaxes the requirements for the power supply system significantly.

4.5 Outlook and Future Work

The presented results of this industrial case study have demonstrated feasibility of a self-sufficient, robust, miniaturized in-tire TPMS module. Based on these results, extended functions as described in Section 4.3 can introduce further added value without creating significant extra system costs, as the vibration scavenger output signal can be reused for deriving complementary physical parameters by means of appropriate algorithms running on an embedded microcontroller.

From a system point of view, TPMS as pioneering reference application for an automotive wireless sensor network (WSN) has high potential to pave the way toward extended intra-vehicle WSN functions. An intra-vehicle WSN could provide a solution for integrating and connecting the growing number of sensor functions within modern vehicles. Particularly non-safety critical sensor functions are suitable candidates for wireless connection. An extended intra-vehicle WSN can

- simplify the ever more and more complex cable tree and therefore
- reduce system costs by eliminating connectors to either or both data communication lines and board battery power system
- contribute to the reduction of overall vehicle weight by omitting cables (in modern cars there are some kilometer cable lengths installed)
- add more flexibility in comparison to a wire-based solution and create added value by making new sensing and monitoring applications possible at locations which are cumbersome or even inaccessible for wire-based access (e.g., at rotating parts and wheels)
- exhibit multicast topology by nature – based on neighborhood – and can therefore provide redundancy and diversity, hence can add reliability – in contrast to wire-based connections – without extra wiring costs

Intra-vehicle WSN technology should not be considered as competitive technology to existing wired field bus systems like CAN, LIN, FlexRay, to name a few; rather the idea is to complement wired solutions and to partially get rid of cables. In particular this includes the ever-growing area of passenger comfort functions, e.g., the penetration of automatic, multi-zone, and high-efficiency climate control systems is increasing and thus the demand for temperature sensors will increase as well. Particularly such "quasi-static" types of sensors with relaxed data rate and update rate demands could be powered by energy harvesters to reach energy autonomy over years, which is a crucial requirement for a future successful automotive WSN roll-out scenario.

4.6 Conclusion

The presented industrial case study has demonstrated the feasibility of a vibration scavenger-driven, highly miniaturized in-tire-mounted TPMS module. By applying careful low power system design and employing robust technology building blocks, proper interconnect technologies and packaging solutions, a volume $< 1\,\text{cm}^3$ for the whole TPMS module, including power supply, at lowest weight could be achieved without compromising performance and reliability issues. The results obtained from this case study can be taken as a cornerstone toward "intelligent tires," which would be able to measure and report additional technical parameters for further enhancement of road safety.

Acknowledgments The presented work has been performed in the course of the European EC-FP6 funded project *e-Cubes*. Web page: http://www.ecubes.org; Contact: Thomas Herndl (thomas.herndl@infineon.com – Graz/Austria) and Werner Weber (werner.weber@infineon.com – Munich/Germany). Special thanks go to Martin Flatscher (Infineon Technologies Austria AG – Graz/Austria), Markus Dielacher (Infineon Technologies Austria AG – Graz/Austria), Josef Prainsack (Infineon Technologies Austria AG – Graz/Austria), Rainer Matischek (Infineon Technologies Austria AG – Graz/Austria), for their excellence in RF transceiver development and demo system integration; Jakob Jongsma (Infineon Technologies Austria AG – Graz/Austria) for his precious technical guidance, Mohammed Abd Allah (Infineon Technologies AG – Munich/Germany) for his highly relevant work on BAW device-level temperature compensation, Horst Theuss (Infineon Technologies AG – Regensburg) for development and design of the packaging solution, and Shi Cheng (Uppsala University – Uppsala/Sweden) for integrated antenna design. Technology development for vertical chip integration, design and implementation of the 3d-Sensor/ASIC stack assembly has been done by Fraunhofer IZM Munich and Berlin, SINTEF, IMEC and SensoNor; particularly many thanks go to Peter Ramm (Fraunhofer IZM – Munich/Germany), Josef Weber (Fraunhofer IZM – Munich/Germany), Jürgen Wolf (Fraunhofer IZM – Berlin/Germany), Maaike M. V. Taklo (SINTEF – Oslo/Norway), Nicolas Lietaer (SINTEF – Oslo/Norway), Eric Beyne (IMEC – Leuven/Belgium), Walter De Raedt (IMEC – Leuven/Belgium), and Terje Kvisteroey (Sensonor – Horten/Norway). Finally, integration of the harvester power supply system would not have been possible without the key components provided by Austrian FIT-IT-funded Project ***PAWiS*** (FFG-No. 810194), conducted by TU Vienna/Institute of Computer Technology and Infineon Technologies Austria AG; Web page: http://www.ict.tuwien.ac.at/pawis Contact: Stefan Mahlknecht (mahlknecht@ict.tuwien.ac.at) Christian Hambeck (hambeck@ict.tuwien.ac.at) and Norwegian RCN Project No. 176485 *Nanomaterials for micro energy harvesting* conducted by Infineon Technologies Sensonor, SINTEF and Vestfold University College Contact: Einar Halvorsen (einar.halvorsen@hive.no).

References

1. Ludvigsen K. (2003) Porsche. Excellence Was Expected, vol. 3. Bentley Publishers, Cambridge, MA, p 1011
2. Otis B et al (June 2004) An ultra-low power MEMS-based two-channel transceiver for wireless sensor network. VLSI Circuits
3. Kowalewski M (August–September 2004) Monitoring and managing tire pressure. IEEE Potentials 23(3):8–10
4. Dubois MA (July 2–4, 2003) Thin film bulk acoustic wave resonators: a technology overview. In: 4th Workshop on MEMS for Millimeterwave Communications, MEMSWAVE 2003, Toulouse, France
5. Dielacher M, Flatscher M, Pribyl W (July 2009) A low noise amplifier with on-chip matching network and integrated bulk acoustic wave resonators for high image rejection. In: Research in Microelectronics and Electronics, 2009. PRIME 2009. Ph.D., Cork, Ireland, pp 172–175
6. Chabloz J et al (February 6–9, 2006) A low power 2.4 GHz CMOS receiver front-end using BAW resonators. In: Solid State Circuits Conference ISSCC 2006. Digest of Technical Papers. IEEE International, San Francisco, CA, pp 1244–1253
7. Flatscher M, Dielacher M, Herndl T, Lentsch T, Matischek R, Prainsack J, Pribyl W, Theuss H, Weber W (February 2009) A robust wireless sensor node for in-tire-pressure monitoring. In: Solid-State Circuits Conference – Digest of Technical Papers, 2009. ISSCC 2009, IEEE International, San Francisco, CA, US, pp 286–287,287a
8. Flatscher M, Dielacher M, Herndl T, Lentsch T, Matischek R, Prainsack J, Pribyl W, Theuss H, Weber W (January 2010) A bulk acoustic wave (baw) based transceiver for an in-tire-pressure monitoring sensor node. Solid-State Circuits, IEEE J. 45:167–177
9. Ramm P, Klumpp A, Weber J, Taklo M (December 2009) 3D System-on-Chip Technologies for More than Moore systems; P. Ramm, A. Klumpp, J. Weber, M. Taklo;

Journal of Microsystem Technologies (Springer), p. 1–5; Published online: Dec 18, 2009 http://www.springerlink.com/content/3n84397t82263112/ DOI 10.1007/s00542-009-0976-1, ISSN: 1432–1858 (Online)

10. Lietaer N et al (October 4–9, 2009) Meet. Abstr. – Electrochem. Soc./MA2009-02/216th ECS Meeting. In: 3D Interconnect Technologies for Advanced MEMS/NEMS Applications – An Invited Talk, Vienna, Austria
11. Schjølberg-Henriksen K et al (2009) Miniaturised sensor node for tire pressure monitoring (e-CUBES). In: Advanced Microsystems for Automotive Applications – Smart Systems for Safety, Sustainability, and Comfort, Meyer G, Valldorf J, Gessner W (eds.). Springer, Berlin, pp 313–332
12. Halvorsen E, Westby ER, Husa S, Vogl A, Østbø NP, Leonov V, Sterken T, Kvisterøy T (June 21–25 2009) An electrostatic energy harvester with electret bias. In: The IEEE 15th International Conference on Solid-State Sensors, Actuators & Microsystems, Transducers, 2009, pp 1381–1384, Denver, CO
13. Lietaer N, Taklo M, Klumpp A, Weber J, Ramm P (November 2009) 3D integration technologies for miniaturized tire pressure monitor systems (TPMS). San Jose, CA, US. In: Proceedings of 42nd International Symposium on Microelectronics – IMAPS 2009

Index

T.J. Kaźmierski, S. Beeby (eds.), *Energy Harvesting Systems*,
DOI 10.1007/978-1-4419-7566-9,

MIX
Papier aus verantwortungsvollen Quellen
Paper from responsible sources
FSC® C105338

If you have any concerns about our products,
you can contact us on
ProductSafety@springernature.com

In case Publisher is established outside the EU,
the EU authorized representative is:
Springer Nature Customer Service Center GmbH
Europaplatz 3, 69115 Heidelberg, Germany

Printed by Libri Plureos GmbH
in Hamburg, Germany